国际科技发展前沿丛书

神经信息工程研究前沿

郑筱祥　主编

高级顾问　潘云鹤　韦　钰　郑南宁
　　　　　吴朝晖　　J.P.唐纳侯
　　　　　J.C.普林希彼

编 委 会　（按姓氏笔画为序）
　　　　　王怡雯　王跃明　刘小峰
　　　　　陈卫东　张韶岷　李　懿
　　　　　杨元魁　杨　怡　赵　挺
　　　　　郑能干

ZHEJIANG UNIVERSITY PRESS
浙江大学出版社

前　言

由中国工程院、国家外国专家局和浙江大学联合举办的"中国工程院首届高层次学术论坛暨神经信息工程前沿研究国际研讨会"受到了许多专家、学者和科研决策者的高度重视。不少世界顶尖大学和研究机构的一流科学家亲临会议，分享当今最前沿的研究进展，发表了远见卓识、启迪思维的演讲，展现了神经信息工程难以估量的巨大发展潜力，将为世界的科学、经济和社会发展带来革命性的改变。与会代表感同身受地领略了神经科学与信息科学交叉融合带来的新境界。

当代科技发展的一个重大趋势，就是生命科学(life sciences)、物理科学(physical sciences)、信息科学(information sciences)与工程科学(engineering sciences)相互渗透与紧密结合，催生越来越多的学科新方向，从而加速科学技术的发展进程，使人们提前享受到科学技术带来的新成果。神经信息工程就是神经学科与信息技术高度交叉融合的综合学科。科学技术发展史也表明，学科的交叉融合导致学科的综合化、整体化发展，神经信息工程就是这种新规律的体现。

本次论坛从"脑机接口及临床应用"、"认知计算与控制"、"神经信息获取、检测与处理"、"神经教育信息工程"和"运动假体神经自主控制"等五个专题阐述重要科学问题，探讨关键技术，总结研究成果，阐述当前热点，展望未来趋势。分享本次会议的成果，相信对广大科技人员和科研决策者具有现实的参考价值，期望能促进同行的交流，进一步推动该领域的发展。

本次论坛的顺利举办和会议论文集的出版，得到了中国工程院信息学部和国际合作局、国家外专局、国家自然科学基金委员会信息学部、西安交通大学、东南大学、浙江大学科学技术研究院的大力支持和帮助，浙江大学求是高等研究院的师生们为此付出了大量的时间和精力，在此表示衷心感谢。

编委会

资助项目

61031002 国家自然科学基金重点项目:植入式脑机接口的信息解析和交互的基础理论与关键技术

2011C14005 浙江省重大科技专项重点国际科技合作研究项目:基于脑机接口的智能假手康复系统研究

60873125 国家自然科学基金面上项目:沉浸式虚拟环境中的脑-机接口技术研究

30800287 国家自然科学基金青年科学基金项目:基于大鼠运动神经编码的脑机接口研究

61001172 国家自然科学基金青年科学基金项目:植入式脑机接口的锋电位点进程估计和动态解析

目　录

一、特邀报告

二、脑机接口及其临床应用专题

三、认知计算与控制专题

四、神经信息获取、检测与处理专题

神经信息工程前沿现状与展望
——2011 神经信息工程研究前沿
国际研讨会综述

郑筱祥

（浙江大学求是高等研究院,杭州,310027,中国）

1 引言

　　神经信息工程是高度交叉的综合性学科,是在神经科学和信息技术的交界面上形成的崭新的多学科交叉的研究领域。神经信息工程包括两个方面:一方面将信息化贯穿于神经科学研究的每一环节,为神经科学的研究提供现代化的高性能信息工具,将不同层次的神经科学研究数据进行分析、处理、整合与建模;另一方面关注如何从神经科学的研究成果中获益,促进信息技术的进步,开发更高级的 IT 产品,以及如何使神经科学更好地利用信息技术来检验模型和假设。神经信息工程主要涉及神经系统信息的产生、编码、存储等过程与机理,以及与人类认知相关的计算、控制和行为感知模型,涉及生命科学、信息科学、工程学等多学科的交叉融合,其发展将推动神经生物学、认知科学、计算机科学、康复医学、微电子学等方面的整体发展,是第三次生命科学革命的重要内容。该领域的研究对理解大脑认知过程、智能信息处理有重要的科学意义,有利于推动高度复杂数据的新型信息感知技术、模式识别技术、集成电路的研究与发展,在挖掘人类认知潜能、研发残障人士的康复设备,以及航天、国家安全等问题上都具有重要的社会意义和广泛的应用前景。

　　近年来,众多国际知名研究机构、高等院校和跨国公司都把神经信息工程作为夺取未来制高点的一个重要阵地,对该领域的研究、发展和人才培养格外重视,期待源源不断地产生新成果,从而推动该领域的迅速发展。在此背景下,中国工程院、国家外国专家局和浙江大学联合举办了"2011 神经信息工程研究前沿国际研讨会",邀请了神经信息工程领域的国际顶尖科学家、国内有较大影响力的知名学者和从事该领域开拓研究的中青年科技人才,着重围绕脑机接口、认知计算和神经教育学等主题进行交流和探讨,分享该领域的最新研究成果,洞察前沿发展动态,聚焦科学问题,探明未来的研究与发展方向。

会议凝聚科学共识,围绕重大科学问题,为该领域的未来发展提出专家咨询意见,绘制发展路线图,实现力量协同攻关,参与广泛而有重点的国际合作,为提高我国社会经济发展和国民健康水平方面做出积极的贡献。

本次会议共邀请了 109 名代表,其中国内代表 87 名,国外代表 22 名,大会主题报告 23 个,其中特邀报告 2 个,涉及神经信息工程的三个重要领域:脑机接口、认知计算和神经教育信息工程,五个专题:脑机接口及临床应用、认知计算与控制、神经信息获取检测与处理、神经教育信息工程、运动假体神经自主控制,会议还特设青年科学家专场。

2　神经信息工程前沿现状

由于神经信息工程所涉及的领域众多,本文仅对脑机接口、认知计算和神经教育信息工程等三个领域中所涉及的脑机接口及临床应用、认知计算与控制、神经信息获取检测与处理、神经教育信息工程、运动假体神经自主控制等五个专题展开讨论。

2.1　脑机接口系统及临床应用

脑机接口(Brain-Machine Interface,BMI)不依赖于常规的脊髓/外周神经肌肉系统,在脑与外部设备之间建立一种新型的信息交流与控制通道,实现了脑与外界的直接交互。自 21 世纪以来,*Nature* 和 *Science* 等报道了一系列脑机接口的重大研究成果,相关研究促进了人们对神经系统的认识,建立了大量复杂信息处理方法,极大地推动了神经、信息与认知等学科的发展。当前,植入式脑机接口已成为国际学术界的热潮,以非人灵长类动物和人类为研究对象已呈趋势,部分成果已应用于临床实践。

美国布朗大学的 Donoghue 教授是植入式脑机接口领域的先驱者之一,他首次在瘫痪病人身上实现植入式脑机接口的临床应用。他提出脑机接口必须解决五大关键科学问题:首要问题是该从哪里获得信号,是运动区还是其他区域? 其次是要知道应采用哪种类型的传感器,颅内传感器还是其他类型的传感器,颅内传感器是否安全、长期和可靠? 第三,从 Spikes 和 LFP 的解码结果中是否能得到足够多的信息? 是否可以从解码结果直接重建手臂的实际运动? 第四,我们需要什么的信号? Spikes 还是 LFP,或者多种类型的信号? 第五,我们需要什么类型的应用? 什么才是有用的设备? 对设备采用什么样的评价标准? 如何评价设备的可用性,可靠性? 是否可以不需要脑外科手术? 他在特邀报告中着重介绍了 BrainGate 技术对上述问题的解决方案及其临床应用,BrainGate 是由美国布朗大学和麻省总医院的团队开发的,并是目前唯一进入早期临床研究的植入式脑机接口设备,可以使瘫痪病人通过自身的神经信号来控制假肢或操纵诸如电脑或服务机器人等外部设备。这些神经技术还将为了解人类大脑生理功能、疾病与损伤机制翻开完全崭新的一页。

亚利桑那州立大学的何际平教授报告了对脑皮层神经信号在不同条件下发生改变或者自适应现象的研究。通过在猴子的脑皮层运动区域(包含前运动区域和感知区域)植入电极,采集经过伸缩抓取训练后的猴子的脑电信号进行分析,验证了灵长类动物大脑中的不同区域通过共同协作而具有学习、主动适应的能力。在临床治疗领域,脊髓损伤(Spinal Cord Injury, SCI)的医疗康复是一个非常重大的课题。SCI 对患者脑部的皮层神经活动及脑电信号的发放都会造成影响。该理论可以帮助研究人员通过分析脑信号找出 SCI 康复的有效治疗及康复方法。

浙江大学郑筱祥教授介绍了浙江大学求是高等研究院的脑机接口研究成果,她领导的团队近年来率先在国内开展植入式脑机接口的研究,实现了啮齿类动物和非人灵长类动物的脑机接口、复杂环境中的动物机器人导航等系统,她从脑机接口、运动神经解码、动物行为诱导、智能控制与多模态反馈等角度阐述科学问题及关键技术,指出协同解码、脑机互适应、多模态反馈、机器智能与生物智能融合等是该领域极其重要的研究课题。

当前,大多数植入式脑机接口系统一般是对运动区神经集群信号进行解码。纽约大学的 Cerf 却另辟蹊径,他报告了采用单神经元解码技术解析人类大脑中与"概念"相关的神经元活动,从而解读大脑的思想,并首次实现了基于高级认知功能的植入式脑机接口。

2.2 神经信息获取、检测与处理

神经信息获取、检测与处理是神经信息工程的重要基础,几年来传感器设计、信号检测及处理方法等发展迅速,极大地推动了神经信息工程的发展。

密歇根大学的 Kipke 教授指出,神经接口的发展正提供越来越强大的包含设计、材料、元件、集成设备的工具包,以推动神经科学的前沿发展以及神经疾病的治疗。除了基于硅衬底的植入式微电极的快速发展,新型的植入式微电极也正在被开发,该类电极利用先进的纳米结构的材料去获得高质量长期的神经记录,同时减少对组织的损伤。另外,多模态的神经电极正在被开发,该类电极能够在神经记录的同时,结合光刺激、神经化学感知以及药物传递的功能。这些技术的发展使得神经电极更为精确,更为可靠,更加能够提取高质量的神经信号。

犹他大学的 Solzbacher 教授报告了植入式微传感器和电极在多通道电

生理和生物标记方面的研究,介绍植入式神经接口和微型生物传感器两种新型的设备和系统,可用于人体和动物神经电生理及新陈代谢等参数的采集,进而推动神经科学的研究以及临床应用。他还详细描述这两种设备的制作和植入过程,以及长期在体和离体使用的性能。

ECoG 电极近年来越来越多地用于脑机接口研究,犹他大学的 Greger 对 ECoG 微电极的优化设计进行了研究,使得大脑皮层表面的空间分辨率能够达到毫米级别。通过植入大脑皮层表面面部运动区和 Wernicke 区的微电极阵列记录到的 ECoG 信号,用于口语词汇分类的研究。

Drexel 大学的 Onaral 教授报告了光学脑成像成果转化的进展,基于近红外光谱技术(NIRS)的光学成像系统是一种广泛应用于脑功能研究的非植入式方法。NIRS 通过检测氧合血红蛋白和脱氧血红蛋白的浓度来间接监测大脑的活动。Drexel 大学的脑光学成像研究团队开发了基于 NIRS 的脑功能监测系统,用以评估健康人和病人的认知活动。该系统具有便携式、安全、价格低和无创等优点,能在多种场合下研究大脑皮层的活动情况,可实现多方面的应用,包括工作绩效评价、麻醉状态监测、神经康复、脑机接口、心理健康治疗等。

佛罗里达大学 Principe 教授阐述了脑机接口研究所面临的巨大挑战,从工程的角度概括了脑机接口系统的几种不同的设计方法、运动皮层和伏核(Nucleus Accumbens)神经信号信息的提取,以及脑机接口系统中动作评价体系的建立。从信号与系统的角度提出了共生(Symbiotic)脑机接口、互适应等课题。

清华大学高上凯教授报告了稳态视觉诱发电位(SSVEP)在脑机接口和认知任务中的应用研究,介绍了各种基于 SSVEP 的脑机接口(BCI),并强调其中的时域、频域、相位、空间的信号分析方法,以及在认知研究中的应用。

2.3　神经控制与运动修复

脑机接口的目的在于实现神经控制、运动功能重建或修复。要实现有效精确的控制,仅依赖解码神经信息得到控制指令是远远不够的,必须综合考虑控制闭环中的每一个因素,包括鲁棒的智能控制、有效的反馈方式、环境感知和上下文关系等要素。

意大利比萨圣·安娜高等研究大学是神经机器人(Neural Robotics)方面的著名研究机构,该校校长 Carrozza 教授报告了智能手(SmartHand)的设计与实验评估研究。手在日常生活中非常重要,由于疾病或事故失去手将会引起一系列生理的和心理的障碍。他们研究并设计了一种用于康复的机械手(SmartHand),可以进行日常生活的抓握、数数和指点,集成了 40 个感受本体和外界的传感器,用于实现自动的控制和特定传入神经的感觉反馈,能够执行控制循环并能与外界环境交互信息。SmartHand 已成为当前世界上最精细的灵巧手之一。

要提高人工假手的功能仍然是一个巨大的挑战,因为在截肢的同时,一部分跟手有关的神经控制信号也失去了。中国科学院深圳先进技术研究院的李光林研究员报告了上臂截肢者对多功能神经假肢的生物控制研究进展。采用目标肌肉神经分布重建(Targeted Muscle Reinnervation,TMR)的外科手术可以把残余的手臂神经转移到可以选择的肌肉位置。在进行了神经移植术之后,这些目标肌肉在皮肤表面产生的肌电信号可以被检测到用来控制假手。为了评估经过 TMR 手术后的上肢截肢病人的表现,他们使用了模式识别的方法对 EMG 信号进行解码和假手动作的控制。

西北大学的 Miller 教授则提供了另一种神经控制方法,他研究如何用脑机接口驱动功能电刺激器(FES)进行辅助抓取。功能电刺激可以刺激支配肌肉活动的神经,使瘫痪的肌肉重新产生动作。目前控制这类抓握动作使用的都是事先制定好的刺激模式,使得手部功能限制在了少数几个事先制定好的模式上。他开发的系统使用猴子运动皮层中记录到的神经信号作为控制信号,为自主控制多个肌肉完成更多任务提供了可能。

纽约州立大学的徐韶华博士报告了适用于神经假体和神经机器人的大脑微刺激方面的研究成果。人类和动物依靠感觉反馈响应外部环境。类似脊髓损伤的神经功能失调会破坏大脑和躯体的联系,导致感觉运动功能的无法恢复的损失。在脑机接口(BMI)方面的最新研究进展表明,通过直接将大脑活动转化成运动命令进而驱动人工器件,对于实现基本的运动功能是可行的。但是,尽管体感反馈对于最佳 BMI 控制是必需的,它尚未被完全地引用到 BMI 中来。他们的工作致力于在大鼠和猴子身上实现将电刺激替代中枢体感通路上的体感反馈。主体感皮层(S1)的多位点记录用于研究前掌的自然触摸和对 VPL 或 S1 的电刺激的神经集群反应。对比结果显示经过参数优化的电刺激可以产生与自然触摸相类似的皮层神经反应。他们也研究了在行为辨别任务中大鼠利用大脑微刺激作为提示的能力。大鼠机器人的研制,是在特定脑区中将电极微刺激替代条件反射式的提示与奖赏。大鼠能够有效地使用大脑微刺激这一概念实现在现实区域中的导航任务。他们的研究提供了实现体感假体的可行性,这需要由心理物理学的研究进一步加以证实。

2.4　认知计算和控制

认知计算的概念兴起于 20 世纪 90 年代,它是用工程化方法重建大脑,开发具有人脑功能的计算机系统,即借鉴神经生物学的研究使计算机具有知觉、感知、认知、思维与意识。研究设计认知计算机的技术关键就是通过对脑结构、动力学、功能和行为的逆向工程设计出具备人类思维能力的智能机器。这将引领一次信息产业的革命性飞跃，将是一次对传统计算模式和计算机体系的颠覆。新的认知计算机将具备自主学习能力等传统计算机难以望其项背的优良特性。同时，如果认知计算机实现了对人脑的模拟，将人脑的认知能力和机器的计算能力完美结合起来，将在众多的应用领域内掀起一场风暴。

布朗大学的 Anderson 教授作了仿脑计算机的研究报告，他认为就短期内认知计算可能最先发展出的主要应用领域为语言理解、互联网搜索、认知数据挖掘和友好的人机接口。

中国科学院上海生命科学院的吴思研究员展示了一些最近在探索 STP 在神经信息处理中的潜在角色研究的相关成果，期望该工作将能为理解大脑怎样处理时态信息打开一扇新的窗口，同时也能有助于开发新的机器学习算法。

西安交通大学的薛建如教授报告了用于智能车辆的视觉认知计算。视觉智能计算是无人车开发中最重要的问题，他所在的团队提出了基于认知模型的智能驾驶系统框架，该框架基于智能体的控制，将系统按照感知决策和控制等功能进行分解，将多传感感知与融合计算和控制计算分离，减少系统计算负担，以提高系统可靠性；针对智能车辆道路和障碍物 3D 重构的需要，提出了车载摄像机外部参数的在线标定三线法；构造了多分辨率车辆检测的假设验证框架；并基于道路边界或标志线构造了参数可变的道路模型及相应道路检测与跟踪算法。

2.5　神经教育信息工程

神经教育信息工程作为神经信息工程与神经教育学交叉融合的前沿学科，其研究内容涉及广泛的工程和教育等多学科领域，其关键的核心技术则主要是信息技术，从信息的获取、信息的处理到信息的解析，从教育方法、教育理念的改进到教育成果的推广应用，信息技术将对神经信息工程研究起到极大的推动作用。另一方面，神经教育信息工程的发展也为信息技术的改进和创新开拓新的天地。

日本工程院院士小泉英明作了"心-脑"科学前沿的特邀报告，指出了神经工程在教育领域的巨大应用潜力和广阔前景。无损脑功能成像技术使得研究可由经验主义转向探知人类复杂的脑功能，大大缩短了以心智为主题的学科和神经科学之间的距离，是进一步推动"Brain-Science & XYZ"跨学科研究的关键因素。他的团队开发出了一种可穿戴式光学成像系统，能够同时观察多个脑的功能和相互作用，而且几乎是实时的。随着"Brain-Science & XYZ"的积累，"应用脑科学"的曙光会在不久的将来出现。

英国 Essex 大学的甘强教授探讨了神经教育信息工程对计算智能的需求,他以在英国的教育神经学研究以及当前教育神经工程研究的不足为切入点,探讨计算智能如何更好地应用于神经科学,包括在高维空间中特征子集的选择,通过机器学习的分类/聚类,以及为可靠的学习困难的早期诊断寻找新的神经记号。

东南大学的俞东川教授和杨元奎教授分别介绍了神经教育工程的进展、社会情绪能力的评估及应用研究。东南大学于 2002 年成立了神经教育学研究中心,研究中心致力于为教育实践和教育政策的制定提供理论基础。目前研究中心已经在社会情绪能力评估方面取得了不少成果。

与会者的报告还涉及诸如 BMI 促成技术(Enabling Technology)、基于功能近红外光谱(fNIR)的脑机接口、光基因学(Optogenetics)、大脑回路映射、运动皮层活动的低维表征等一系列重要的研究成果。限于篇幅,本文不再一一赘述。

3　总结与展望

与会科学家分别从各自研究工作视角出发,介绍了神经信息工程前沿发展动态,分享最新最前沿的研究成果,并对未来学科发展方向提出了自己的独到见解。得益于神经科学、信息科学与工程科学的交叉融合,神经信息工程的研究与发展呈现加速的势头,在深度上向终结科学问题不断逼近,在广度上向其他学科和应用领域快速渗透和扩散。我们看到,脑机接口、认知计算和神经教育信息工程这三个具有代表性的既互相支持又互相渗透的研究领域,在已经茁壮成长的神经信息工程这棵参天大树上,绽放出令人惊异的绚丽之花,并终将结出累累果实。展望未来,我们需要从不同层次、不同粒度对脑信号进行分析、建模、挖掘和利用,针对不同需求构建应用系统,解决并突破神经信号处理、认知计算与控制、环境智能、大脑可塑性、脑机互适应、生物智能与机器智能的融合等技术难题,有望在脑机接口、认知计算、神经教育等方面研发应用产品,更好地造福人类社会。我们认为未来我们还需在以下几个方面进行重点研究:

3.1　植入式脑机接口

瞄准国际神经科学与信息科学的交叉前沿和我国医学康复和医疗产业的重大需求,发展可应用于在体动物或人体的多个层次神经信息检测、分析和处理的新技术和新方法,从细胞、组织和整体等多个水平实现机器对于神经活动的信息解析、运动表征和行为感知,构建基于双向信息传递的脑机接口技术平台,开发若干可用于残障人群恢复感觉和运动功能的辅助装置和系统,重点针对人口老龄化带来的运动功能障碍问题,结合先进的神经信息工程技术研发适合国情的多反馈闭环交互式复杂运动功能康复的辅助运动康复机械设备和功能性电刺激设备,最终实现神经信息技术的快速和跨越式发展。针对城市安全、救援与反恐等国防与公共安全重大需求,综合运用生物电子、生物微机电系统、神经接口等生物体与电子接口技术发展生物-机电复合型智能机器人,融合动物感知、动物行为调控与机器人学的新技术,在安全与搜救方面取得示范应用。

3.2　认知计算

人和动物在面对和处理未知环境中的问题时表现出超强的鲁棒性。通过增强智能系统的这种类人的感知能力,开发与构建一种机器人系统或其他人工认知系统。该系统可以处理与解释多种感知信息(图像、语音等),并在动态实际环境中具备灵活的决策能力,自主完成任务目标。借鉴生物科学理论,在感知、理解、交互、学习与知识表达方面产生创新性方法与理论。重点突破基于脑认知的视觉加工模型,发展基于视觉的自然环境感知技术,基于认知计算,实现多传感器跨模态跨尺度信息融合,能模仿人体视觉系统的神经网快速地识别自己周围的世界。开发基于人类视觉系统的超级计算芯片,并形成典型示范应用工程,如自然环境中的机器人系统、复杂系统中的感知与控制等。

3.3　神经教育信息工程

对神经教育信息工程的研究将从神经教育学以及神经信息工程的整体国际态势出发,涵盖神经教育信息工程科学研究层面、社会服务层面以及政策研究层面,开展儿童个性化多媒体数据采集、脑电信号、电生理信号采集,社会信号处理技术、脑电分析技术、脑机接口技术、基于大规模多媒体数据的有效储存和智能挖掘技术等对我国信息技术及相关领域可能带来的融合和跨越式发展的关键科学问题和技术热点进行组织研究。

3.4　人工智能与生物智能的融合

大脑具有一种惊人的将跨意识的多重含义信息整合的能力,它可以毫不费力地创建时间、空间和物体的种类,以及探知感官数据的相互关系。大脑可以完成各种无以伦比的技艺,令现在的计算机望尘莫及。而现有计算机信息系统(或人工智能系统)在计算能力方面有着无与伦比的优势。一方面需要研究设计认知计算机,通过对脑结构、动力学、功能和行为的逆向工程设计出具备人类思维能力的智能机器。更重要的是要研究如何将人工智能系统与生物智能系统融合,使两者互为适应、协同工作,实现人脑的认知能力和机器的计算能力的完美结合。

　　神经信息工程是一门交叉学科,涉及神经科学、计算机科学、数学、临床医学和各类工程技术,要求相关学科的研究人员共同协作,探索并解决其中的关键和基础问题,甚至要全球范围内开展全方位合作,共创共享人类知识资源。本次会议也将促进国际上各个团队之间的合作,加速神经信息工程的研究,同时也为我国的神经信息工程的发展规划起到较好的指导作用。

一、特邀报告

利用意念控制机器
——神经工程技术在运动功能重建和脑伤治疗方面的未来趋势

John P. Donoghue

（布朗大学布朗脑科学研究院，Providence，02912 USA）

摘要：神经工程技术作为近十年来最活跃的研究领域，已经推动了多种新型医疗仪器的发展，并在神经损伤功能的修复和重建及神经系统疾病的诊断和治疗方面展示出了巨大的潜力。当前，基于神经电刺激技术的人工耳蜗和深部脑刺激技术分别在听力恢复和帕金森病的治疗方面取得了成功。这些植入式"脑机接口"，以及在视觉假体等其他方面应用的神经刺激器已经对成千上万人的生活产生了重大的影响。另一方面，作为一种新的神经工程技术，能"感应"神经信号的神经接口系统也已进入了早期的临床实验。这类脑机接口旨在帮助瘫痪病人重建一条新的独立交流与控制的通路。BrainGate 是由我所在的美国布朗大学和麻省总医院联合开发，且是目前唯一进入早期临床研究的植入式脑机接口设备。该设备可使瘫痪病人控制假肢及操纵电脑或机器人等外部设备，其控制信号主要来自大脑运动皮层中负责手部运动脑区的神经信号，包括了神经元的发放电和场电位信号。这些信号可通过植入皮层的一个 100 通道的慢性微电极阵列记录得到，并用于提供各种运动指令。对5 个瘫痪病人进行的早期研究结果充分表明了大脑皮层中与手部运动相关的脑区在损伤后的很长时间仍可持续记录得到神经元活动。被试者能够控制光标的移动或者是操纵机械臂抓取物体。由于功能性电刺激系统已应用于瘫痪病人群体，因此脑机接口的另一个主要目标就是将运动命令连接到植入式的功能性电刺激，使肌肉得以重新活动。早期的无线植入式系统目前正在进行初步的临床前测试，相关研究表明高频带带宽的植入式系统也有很高的可行性。除了应用于脑机接口，神经工程技术还为观察人类大脑活动提供了一个具有很高时空分辨率的监测工具。总而言之，由于神经工程技术能在细胞水平记录多个单神经元活动和场电位信号，这将为进一步了解人类大脑的生理功能、疾病与损伤的机制等翻开崭新的一页。

关键词：神经工程技术；脑机接口这；BrainGate；神经传感器；功能重建

Abstract: Neuro-technology is an emerging field creating a variety of new devices with the potential restore lost functions as well as diagnose and treat nervous systems disorders. Currently available neurotechnologies based on stimulating the brain including cochlear implants to restore hearing and deep brain stimulators (DBS) for Parkinson's disease. These implanted "brain interfaces", already in use by tens of thousands of humans have dramatic effects on quality of life. Stimulation devices are also being developed for vision prostheses and other applications. Neural interface systems that sense neural signals are a new neuro-technology, now in early-stage human trials. These brain-machine interfaces (BMI) aim to restore the lost function of paralysis people. BrainGate is a unique BMI in early stage human clinical trials being developed by our group at Brown University and Massachusetts General Hospital. BrainGate is being developed to allow paralyzed humans to use neural signals ordinarily to control the arm to operate devices such as computer software or robotic assistants. Control signals are derived from the neural activity in arm area of motor cortex, through a 100 chronically implanted microelectrode array that enables recordings of action and field potentials, both of which can provided useful command signals. Preliminary experience from five tetraplegia participants, including one for more than 5 years, demonstrate that arm-related motor cortical signals remain in motor cortex long after injury or disease onset and can record for extended durations. Participants are able to perform point and click actions of a computer cursor or move a robotic arm to grasp objects. A major goal is to reanimate muscles by connecting motor output to implanted functional electrical stimulators, which has been demonstrated by a person with tetraplegia using a simulated FES system. Early stage wireless, implantable systems now in initial preclinical testing also suggest that high bandwidth implantable systems are feasible. Beyond BMI applications, the same technology may provide new approaches to monitor human brain activity patterns at a fine temporal and spatial scale. Finally, the ability to sense at the level of multiple single neurons and local field potentials opens up an entirely new perspective on the function of the human brain as well as the impact of disease and injury.

Keywords: Neuro-technology, Brain-machine interface, BrainGate, Neural sensor, Functional restoration

1　引言

1.1　即将到来的神经工程时代

非常感谢大会的组织者,提供了这样一个机会,可以和大家共同探讨神经工程技术在运动功能重建和脑损伤治疗方面的研究、应用及未来发展趋势。毫无疑问,人类将迎来一个激动人心的神经工程时代,而我们恰好有幸成为了这项技术的先行者。借助神经工程技术,人类将更有能力对各种神经疾病进行诊断和治疗,甚至能重建受损的神经功能。目前在受损听力功能的恢复中已应用了该项技术,且正在开展视觉功能修复和瘫痪病人运动功能重建的研究。人们对心脏起搏器的印象源于该技术出现的 20 世纪 50 年代,在经历半个多世纪的发展之后,它已成为一个完全植入的医疗器件,并成了常规手术治疗手段。尽管目前神经工程技术的发展才刚起步,距离广泛应用仍有很长的路要走,但可以预见,在不远的将来,神经工程技术将进入到我们的日常生活,甚至与生活和工作密不可分。

1.2　神经接口的分类、研究现状和进展

根据大脑与外部设备间信息传输方向的不同,目前的神经工程技术主要分为两大类,输入型(Write-in)和输出型(Read-out)神经接口。对绝大多数神经工程研究者而言,最终目的是将这两类神经接口整合到一个系统中去。当前输入型接口主要利用电刺激将外界的信息输入大脑,用于恢复受损的感觉信息或者用于神经调控。本次报告的主题是关于从大脑内读取神经信息,用于受损功能的恢复,时间允许的话还将涉及该技术在大脑状态测试和评估方面的应用,也就是特别强调的大脑与外部世界间连接的稳定性。

当前的植入电子耳蜗技术已相当成熟,全世界范围内已有超过 17 万人通过该项技术实现了听觉功能的恢复及正常的交谈。近年来,还有少数人通过植入眼中的电子刺激器帮助恢复视觉,只是现在的视觉假体技术才刚起步,远不如电子耳蜗技术成熟,只能提供非常简单粗糙的影像,且离看到的真实图像还存在非常大的差距。另一方面,深部脑刺激作为一种输入型脑机接口技术,近年来对其进行的研究取得了较大的进展。在美国,每年新增大约 5 万例帕金森综合征患者,由疾病所导致的运动障碍、震颤和肌肉僵直等症状对他们的生活和工作产生了极大的影响。深部脑刺激技术通过植入脑内特定核团的电极施加刺激,能够有效地缓解上述症状。这些例子都很好地表明了新的神经工程技术可有效提高生活质量,到目前为止,已有大约 5 万人因该项技术而受益。因此,在脑内植入器件已不再是一个神奇

的想法,而是一项实实在在的技术,已经应用于临床医学,也有了完整的危险评估标准。

1.3　输出型脑机接口研究的主要用途和面临的挑战

本次报告的主题是关于输出型神经接口,即从大脑中读取神经信号,以及与此相关的一个名为"BrainGate"的项目,希望能够借此帮助到瘫痪病人。瘫痪是由多种疾病造成的,可以是任何一种形式的大脑与肌肉运动系统之间的通讯扰乱,比如大脑和脊髓之间的通路或脊髓和肌肉系统之间的通路发生了中断。因为大脑无法继续控制运动肌系统,截肢也可看作是一种神经通路的中断。根据统计,美国有560万不同程度的瘫痪病人,其中小部分人的症状十分严重,除了眼睛外,几乎无法控制身体的其他任何部分。

脑机接口(Brain-Computer Interface 或 Brain-Machine Interface)主要实现了大脑和外部世界的直接连接(如图 1 所示),如控制计算机、辅助技术、机器人、人工假肢甚至被试者自身的肌肉。脑机接口所面临的挑战也是本次报告的主要内容之一。由于神经工程是一门交叉学科,因此挑战也涉及了神经科学、计算机科学、数学、临床医学和各类工程技术等。神经工程领域的研究要求相关学科的研究人员共同协作,探索并解决其中的基础和关键问题,如从哪个脑区提取神经信号,分析哪些神经信号,利用何种传感器来稳定并长期地获取信号,怎样将神经信号转换为控制指令以及需要哪些技术真正实现目标设计所需要的功能等。这几个重要的关键问题需要大家一起进行探讨。

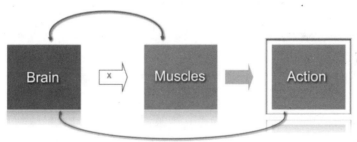

图 1　脑机接口系统框图

1.4　脑机接口的目的

脑机接口研究的最终目的是实现安全有效的控制,进而为瘫痪病人与外界进行交互提供一种新的手段。当前该技术所关心的主要问题有:如何采集信号,是通过植入式(颅内记录)还是非植入式(头皮上记录)记录神经信号?传感器是否可见?控制的水平是一维的还是多维的?脑机接口可连接怎样的辅助设备,是完成简单的开关、控制一台电脑还是实现对自身肌肉系统的控制? 这些都是在研究时所必须考虑的,只有这样才能明确目标所在,清楚在以后的研究中需要完成的工作。

2　BrainGate 项目研究

BrainGate 是一个较早的脑机接口项目,虽然无法帮助解决所有的问题,但还是可以通过它对一些重要的问题进行研究,也取得了一些成果。该项目的目标是在大脑与外部世界间重新建立一个连接,即通过重建上肢和手的运动实现对外部设备(机器或肢体)的控制,所要解决的关键问题有以下几点:

- 使用安全性;
- 提供上肢或手的功能;
- 提供自然控制,无需过多注意;
- 方便获得;
- 超过 10 年的长期稳定性;
- 便于携带;
- 符合美学要求,尽可能小,最好不可见。

下面介绍 BrainGate 项目早期的临床实验结果,以及由此所带来的一些争议,还包括了该项目的最新进展以及对于神经接口未来发展趋势的个人看法。

2.1　BrainGate 系统构成

BrainGate 系统中首先包括了植入到大脑初级运动皮层区控制手的部位的一个包含 100 个通道的 Utah 电极(4mm×4mm),电极通过电缆连接到头骨的基座上(1 欧元硬币大小),用一整套放大器和记录设备记录到 96 通道的神经信号。这些信号包括场电位信号和 Spike 信号,利用信号处理器对以上神经信号进行解码,转换为控制命令。当被试者想象运动鼠标或者运动手臂时,解码得到的控制命令就可用于操作电脑或者玩视频游戏等。

2.2　初步临床实验

BrainGate 项目已经过美国 FDA 批准,临床研究数据源于我们的实验室。从 2004 年开始到现在,已有 5 位被试接受了电极植入,并进行了相关实验。这 5 人均为重症瘫痪病人,无法控制四肢的活动,其中两个为中枢脊椎损伤,一个为脑干中风,另外两个为肌萎缩侧索硬化症(Amyotrophic Lateral Sclerosis,ALS)。另外,在已得到批准可开展的新临床实验中,目前有 14 名志愿者计划参与。本报告中的数据主要来自被试者 S1 和 S3。

2.2.1　记录电极和电极植入

所有的被试都使用了相同的微电极阵列,由 100 个微电极组成,大小为 4mm×4mm,植入到初级运动皮层的手臂和手的功能区,功能区已预先通过磁共振成像进行了定位。这里用一段视频,可展示电极阵列植入的全过程。借助一个医用植入器,通过其顶部的气锤可将电极快速打入到皮层中去,在这个过程中需要注意的是,击打应该在脑膜鼓起的时候进行。

2.2.2　稳定的神经信号

项目研究中,最重要的是保证信号的稳定性。在所有的 5 个被试中,用于传输运动皮层神经元信号的神经通路已中断了好几年,尤其是一个病人自从瘫痪后,已有九年无法控制上肢。可以想见的是,该区域的神经信号可能已经没有了,或者即使有神经信号保留下来,也可能已经同手的运动无关了。然而,皮层记录到了神经元的活动,且与手的活动密切相关。在之后的实验中,实验员让被试 S3 想象手的张开和闭合,可以发现,当被试想象手闭合时,神经元处于静息状态不发放,而当被试想象手张开时,神经元处于活动状态。此结果充分说明了初级运动皮层的神经元仍然处于活动状态,且和神经通路正常的时候一样,被试仅需想象一下手的运动就足以引起神经元的活动。这也是我们研究的基础,实验结果表明,我们仍然可以利用神经元的活动来获取被试的运动意愿,重建与外部世界间的通讯。

更为重要的是,多年来基于非人灵长类动物猴子的研究表明猴子初级运动皮层的神经元还可直接调制运动的方向。图 2 的结果表明,当被试想象手

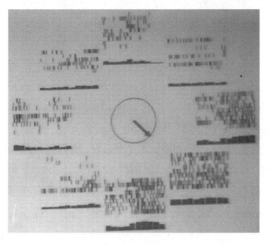

图 2　初级运动皮层神经元表现出的方向偏好性

在八个方向上运动时,初级运动皮层神经元表现出了不同的方向偏好性,这些信息可用于校正神经活动与想象运动方向间的映射关系。这是一段早期临床

研究的视频。被试坐在轮椅上玩视频游戏。由于高位截瘫的缘故，被试只能用其大脑活动，主要从神经元的动作电位获得的信息控制屏幕上的光标。经过若干次训练之后，被试就可以用神经信号成功控制光标画圆。此外，被试还可用神经信号控制假肢手的张合等。2006 年发表在 *Nature* 上的视频展示的是利用大脑神经信号直接控制一个玩具机械手，被试通过对屏幕上光标的运动控制，可用很简单的方式抓住一些小糖果，并将这些小糖果分送给别人。这是帮助被试实现和外部世界通信的方式之一，在经过若干年训练之后，可控制简单的机器人。

2.3　临床实验展示

这些早期的工作表明了以下几点：神经外科手术过程非常简单；手臂的信号在瘫痪多年后仍然存在；无需学习，仅通过校正滤波器（filter）就可直接获得关于运动的意愿信息；无需特别的注意力要求；可实现连续控制；脑机接口可实现对各种外部设备的控制。

3　植入式脑机接口研究中的关键科学问题

伴随早期脑机接口研究工作而来的是各种各样的争议。这些争议问题非常重要，需要仔细加以考虑。在 BrainGate 项目的实践中可以得到关于这些问题的答案。上述各项工作主要基于颅内记录的脑机接口，这也是本次报告所主要关注的对象，其所涉及的关键问题包括：

（1）需要从哪里获得信号，运动区还是其他区域？

（2）需要何种类型的传感器，颅内传感器还是其他类型的传感器？颅内传感器是否安全、长期和可靠？

（3）是否可以从 Spike 和 LFP 的解码结果中得到足够多的信息？是否可以从解码结果直接复制手臂的实际运动？

（4）需要何种类型信号？Spike 或者 LFP，或者多种类型的信号？

（5）可进行何种类型的应用？有用的设备是什么？采用怎样的标准对设备进行可用性和可靠性的评价？应用是否不需要脑外科手术就可以实现？

所有这些挑战和争议都需要进行充分的讨论。下面是针对每个问题所给出的观点。

3.1　脑区：负责抓取的皮层

第一，关于获取信号的区域选择问题。无论是在猴子还是人，除了已选择的运动皮层区域,还有其他大量可供选择的区域与手臂相关,有些是直接相关,有些是部分相关。这些都是可以获得相关信号的潜在区域。在 19 世纪 70 年代就已经发现,主运动皮层(Primary Motor Cortex)是和手臂运动相关的大脑皮层区域。经过 100 多年的发展,人类已经可以知道手臂运动功能具体的定位,获得了大量关于运动的生理医学数据,根据从这些区域获得的数据,可帮助那些手臂无法运动或是失去手臂的人。相比较其他区域,这种类型的控制更为自然,这是做出所有决策的基础,虽然不见得是最正确的选择,手臂运动区域仍然是一个好的选择。

3.2　传感器:植入式电极阵列

第二,关于颅内传感器是否安全可靠的问题。选择获得大脑活动信号,可以有多种方式和途径,而电极的选择很大程度上取决于想要获得什么类型的信号,如需获得锋电位(Spike)信号,就要穿透大脑皮层区域,采用植入式电极,也可以有其他方式。2009 年发表在 *Neuron* 上的综述文章,有关于这部分内容的详细讨论。之所以选择这种植入式电极,是因为它可以同时提供更多的信息,是可选方式中最适合的。同时,也希望电极技术可以提供多种不同类型的信号。图 3 显示的是可以从大脑中获得的两类最主要的信号,一类为局部场电位(LFP),主要是突触电流引起的一种慢波信号;另一类电信号为锋电位,是记录到神经元的动作电位信号。这两类信号其实是同一个信号经过不同的滤波器获得的。这两类神经电信号都来源于大脑,因此都可以用于检查和分析。在临床中,了解更多使用更多的是锋电位信号,也是今天报告中所涉及的对象。而 LFP 也能很方便地用同样的方法得到,其中也包含了大量的信息。

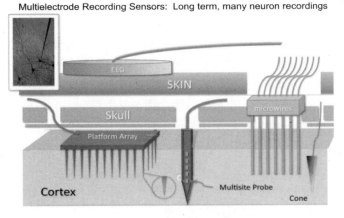

图 3　各类神经信号传感器

3.2.1　Utah 微电极阵列

前面实验中所选用的 100 通道的 Utah 微电极阵列,是一个非常精美的技术产物。在 20 世纪 60 年代,当时的研究发现,当金属电极的顶端被制作成尖端时,记录效果会非常好,虽然时至今日对于其中的原理仍知之甚少,但其确实能记录到非常好的信号。与以前的电极不同,现在的电极材料既不是金属也不是铂金,而是用硅通过蚀刻等一系列特殊的工艺制作成的。Normman 教授先后为这个微电极阵列设计制造了多个型号。作为一项开创性的工作,该电极取得了相当大的成功。

3.2.2　植入式电极的安全性

对于各类记录信号的传感器,需要考虑众多相关因素,尤其针对植入式传感器,因为有手术的介入,安全非常重要。手术总是伴随着风险,在植入过程中由于血管被穿破导致出血,可能引起进一步的感染。其他原因还包括了植入引起的神经元损伤和死亡,以及大脑的重组。在实验中也出现了各种不同的情况,最后导致了电极阵列无法记录到信号,出现问题的时间也不尽相同,有三个月的,有六个月的,有甚至一年之后的。

3.2.3　植入式电极的主要失效原因

在猴子实验方面,目前尚无数据说明哪些因素会影响一个特殊电极阵列的使用寿命。正在开展的临床实验研究,还没有得出确切的结论,而在早期的临床研究中,却也发现了一些非常鼓舞人心的实验结果。由于时间的关系,这些结果将进行快速的展示。根据实验结果,电极失效的原因可分为三类:第一类是生物反应问题。实验中所采用的电极阵列为 Utah Array,被放置在皮层的表面,是最容易导致电极失效的因素,任何引起脑内血管出血的因素都可能导致信号的消失。根据植入电极在脑中的显微示意图所示,由于脑内具有丰富的毛细血管,电极植入无可避免地将引起血管损伤,同时也可能引起电极周围神经细胞的损伤。此外,电极周围的免疫反应,以及电极周围的胶质细胞生长等一起构成了非常主要的生物反应问题。第二类是材料问题。由于受到生物体的排异反应,系统中所用材料也可能导致电极无法工作。有些材料的生物相容性非常好,可持续工作很长时间。而如果电极周围的包裹层破裂,会导致生物组织中的液体渗透,从而破坏电极和包裹物间的连接,进一步破坏电极外部的保护涂层。由于材料的连接部分往往非常复杂,两个材料间的接口往往会因为容易损伤而无法工作。第三类是机械问题,主要是牵拉。大脑运动和组织反应引起的牵拉可能导致连接脱落。实验中,有很多失效都由牵拉引起。而实验过程中大部分失效都与生物反应相关,因此可以认为在三类失效因素中生物反应最为重要,至少实验所采用的 Utah 电极阵列是如此。

3.2.4　植入式电极的长效性

结合所做的组织学实验及得到的实验结果,这类电极最重要的特征在于其长效性。得益于强大的技术,即使脑组织发生了生物反应,电极仍能持续很长时间发挥其作用,而无需考虑安全性问题。但这个电极阵列,仅针对非人灵长类或者人类实验所采取的方法。在这方面关于脑组织反应的问题,所有的这些因素都非常重要。

2005 年,在研究的早期所发表的关于电极长效性的文章中,对 3 个猴子脑中的电极阵列进行了跟踪,分别记录了 514,154 和 83 天。虽然在写文章时因为觉得电极信号的时间点到了没有继续跟踪,实际上微电极阵列在一年后仍然有效地记录信号。

3.2.4.1　动物实验研究

从已经埋植的 71 个电极中获得了大量的实验数据。图 4 展示的是在八只猴子上记录了 5 年的电极情况。黑白方格部分(ended)表明部分电极出了些问题;对正常的电极则进行有选择性的记录后便停止,电极没有出现任何故障;波纹形状部分(tech)表示由于技术原因引起的失效,这些原因主要来自猴子,比如猴子很容易将电极的连接口弄坏,或是将牙科水泥弄松,而这些原因不会出现在人身上。其他的一些故障主要与中枢神经系统以及外科医生的手术过程有关。

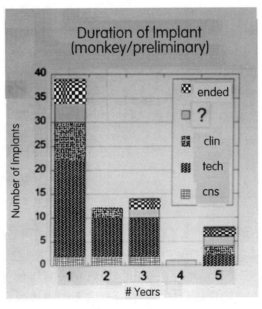

图 4　猴子电极植入持续时间

3.2.4.2 临床实验研究

更重要的是来自临床实验上的成功。从来自五位临床被试者的实验数据可以看到都有多年的神经记录信号，其中一个的记录时间已超过五年，并依然在参与实验。如表 1 所示，前两个有些问题，一个是因为材料引起泄漏而不得不修理，另一个则是因为连接器故障而不得不停止实验。还有两个 ALS 病人由于患了其他疾病而不得不终止实验。对这些患者的记录都不足一年，但不是因为技术上的原因。

临床实验中最重要的是阵列功能保持良好。从给出的被试 S3 在电极植入超过 1000 天时的录像可以看到，她需要用神经信号控制光标移动到目标位置，并且还需要通过想象握紧手来实现点击动作。在《神经工程杂志》(*Journal of Neural Engineering*)即将要刊出的文章中还包括了用更小的光标和随机目标进行尝试的实验。文章主要讨论的是 S3 在植入电极 1000 天后的性能，第一天有 100%，接着几天稍差，随后又会升高。该被试的的实验非常成功，表明其有能力操控该项技术。这也是表明植入式脑机接口技术可长时间使用的明确证据。

表 1　人的电极植入状态表

Study participant	Days array	Status	Sensor Issues (preliminary)
S1 24yo M C4 ASIA A	1.33 y	Completed	Material (leak)
S2：53yo M C4 ASIA A	2.16 y	Completed	Material (connector)
S3：55yo F Brainstem stroke	5.27 y	Ongoing	Tethering (part recovered)
A1：37 yo M ALS	296 d	Deceased	None
T1：48 yo F ALS	307 d	Deceased	None

此外，为改进解码效果引入了适应性解码算法，同样是 S3 的录像，在植入微电极阵列五年后，做同样的"Point-Click"实验，其操控准确率依然能保持原有水平。虽然 S3 是目前唯一一个电极植入超过五年的被试，该实验证据很好地说明了植入式脑机接口技术会在今后若干年内持续应用并发展。因此根据五年来的临床实验情况，该技术具有非常稳定的安全性，但由于目前仅有一个病人，尚需要更多的数据支持，让阵列在人身上安全工作十年还是很有

希望的。

3.3　基于 Spike 的解码

这里的关键问题是可以从单个 MI 中解码得到多少维度的信息。实验证明，可以获得的信息有很多。在猴子实验中，可训练猴子根据提示如立方体，抓取不同的物体，在记录 MI 信号的同时，结合运动捕获设备记录猴子的动作，从中希望知道可记录多少种不同的动作，而根据记录到的信号又能重构出多少种动作。

视频演示了重建整个"抓"动作的过程。即使是很小的 MI 区也有能力重建整个"抓"的动作，动作包括了 10 个自由度，分别是肩关节、肘关节和手，阴影表示真实手的动作，实线表示了预测的动作。虽然动作并不完美，但事实上，若能在人身上用运动皮层上的单一电极阵列记录到同样的信号，就可解码出整个"伸—抓"动作。

3.4　神经信号

除神经元活动外，场电位(FP)也能作为一种可选信号用于神经解码。对运动准备和运动执行，LFP 表现出了不同的特性，从能量分布上看，在运动时，低频和高频成分表现更好，而在运动准备过程中，中频(大约 $10\sim50\text{Hz}$)能量较其他频率更好。此外，在运动执行前，慢波会表现出一个明显的负峰，而且运动方向和 FP 信号间存在着一定的相关性。

对三维空间上手的运动轨迹重建，分别利用了高频 FP 信号和低频 FP 信号对猴子上肢(包括手指)的运动参数进行解码，同时也正在尝试将两类 LFP 信号相结合应用于运动的神经解码过程。

3.5　BrainGate 的应用

这部分将着重关注神经接口技术的应用，当前利用该技术进行的系统仪器方面的开发，希望能够帮助到如瘫痪病人之类的特殊人群，相关工作主要包括：通讯界面，面向无法说话的病人；辅助系统，面向瘫痪病人；神经假肢，具有七个以上自由度，可用大脑直接控制；功能性电刺激，与其他单位(Case Western Reserve University)合作，试着将 BrainGate 的信号给肌肉电刺激系统。

以下将介绍两个最近研发的实例，虚拟键盘打字和机械手控制，这两者都处于早期阶段。

3.5.1　虚拟键盘打字

同样是被试 S3，在植入 Utah 电极阵列五年后，可用自行开发的打字系统进行演示，系统中所使用的是标准美国键盘。首先，将光标移到字母上并选定,当备选的单词会像智能手机一样出现在屏幕右侧时,就可选中可能的词。

根据上述原型系统又设计了呈放射形的打字机,与手机键盘类似,被试可先选字母,然后从备选单词中选择单词,这样能进一步提高打字的速度。目前S3每分钟能打20个字母。打第二个字时,为她展示要选的字母,通过点击会出现更多的备选单词,这样就能在屏幕上打出所要的单词。借助这样的系统,被试既能打字,又能移动鼠标,也可用于其他打字接口应用中。

3.5.2　机械手控制

供瘫痪患者使用的辅助机器人正开始进行相关的测试。健康人用鼠标控制机械臂没有任何神奇的地方,而计算机可通过控制鼠标移动机器手臂,使之能抓起杯子,并给被试喂水喝。

这里还是被试S3,已通过BrainGate连接上机械假肢,在经过校正过滤后,就能用神经信号实现控制机械假肢。在早期阶段,患者能很快实现对机械手臂一维的准确控制,即简单地上下移动机械手。

二维控制实验可称作是一个桌面版本的Center-out实验,就如经常在猴子实验中所看到的,这是一个很好的例子。桌面上有一个酒杯,要求被试将酒杯从桌子上举起,放到中间一个目标上。这里,机械手在桌面上所有的水平移动以及闭合都是通过神经信号控制的,而机械臂的上下移动则不是。到目前为止,已可以实现很好的二维控制。三维运动控制的实验刚开始,效果不好,需要考虑进一步改进以实现更好的三维控制。

3.6　研究进展总结

对现有脑机接口研究进展进行总结,包括如下部分:

(1)记录区域:对解码上肢运动,初级运动皮层的手臂区域是一个好的脑区,其他脑区也是可选的;

(2)记录传感器:对于传感器的稳定性和安全性,目前的临床证据还较为有限,但因为已有一个超过五年的临床案例,植入式传感器是非常有前途的。同时,其他实验室已有报道表明,在动物实验中(猴子)可记录的时间更长;

(3)神经解码:虽然目前临床上仍在研究人的神经信号与上肢三维活动间的关系,在二维控制方面的研究结果已令人非常鼓舞;

(4)神经信号:动物实验和临床实验表明,可以从Spike中提取到大量与运动相关的信息,不同的LFP信号的不同频带中也同样包含了多种运动信息;

(5)应用:本报告中所展示的一些仪器和系统设备暂时还无法提供给所有疾病患者,也不能说这些仪器对患者的日常生活都有用,但是相信将来这肯定能对患者的生活产生重要的影响。

4　下一代 BrainGate 和植入式脑机接口的发展

最后,用两分钟时间对脑机接口技术的研究做一个总结和展望。BrainGate的目标是要开发一个无线的、完全植入的系统。为了达到这个目的,还需要开发一系列可用的新技术和设备,实现对目前部分功能的代替或升级,包括完全植入,无需头颅上的连接口;便携式,无需一大堆电子设备;自校正,无需专业技术员的操作。只有这样的 BrainGate,才能完全为患者提供所需。

图 5 是正在研发的下一代 BrainGate 系统原型框图,外部仅需一个类似 iPhone 大小的神经信号处理器,内部则采用完全植入式的记录电极,在头骨外有一些电子系统,通过无线传输的方式经过皮肤将神经信号传输出来,在外部处理器中对信号进行分析和处理。

团队里的工程专家 Arto Nurmiko 已领导开发了第一代原型——布朗植入芯片,可利用光学方式将神经信号传出皮肤。神经芯片的三个主要部分:传感器、传输电缆和无线电子设备,都埋植在头骨的顶端。该神经芯片目前已在 2 只猴子上开展了预实验,记录到了它们的神经信号,有些最新成果已发表在文章中。

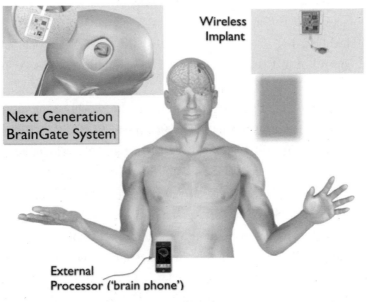

图 5　下一代 BrainGate

5　神经工程是交叉学科

在神经工程领域,可实现很多有希望的目标,也可以采用多种技术路线。如之前讨论过的刺激方法,以及利用记录技术恢复运动,记录技术还可用于评估模块性,预测疾病,帮助治疗严重生理疾病患者。同样重要的还有可利用该手段学习关于大脑的精确的基础的知识,这包含了非常大量的信息,且是全新的,可以帮助人类考虑大脑是怎样工作的。

最后,需要强调的是,神经工程是一个交叉学科,我们的团队中包括了前面提到的来自工程学科的 Arto,计算机科学的 Michael,临床医生 Leigh,开展了很多临床实验,我来自神经科学,仅是这个交叉团队中的一员,是很小的一分子。正是依靠团队的力量才能让这一切发生,使意念控制成为现实,才有了现在的成绩,在这里再次向我的团队表示感谢。

参考文献

[1] Serruya MD, Hatsopoulos NG, Paninski L, Fellows MR, and Donoghue JP. Instant neural control of a movement signal. *Nature*, 2002, 416(6877): 141 - 142

[2] Hochberg LR and Donoghue JP. Sensors for brain-computer interfaces. *IEEE Eng Med Biol Mag*, 2006, 25(5): 32 - 38

[3] Hochberg LR, Serruya MD, Friehs GM, *et al*. Neuronal ensemble control of prosthetic devices by a human with tetraplegia. *Nature*, 2006, 442(7099): 164 - 171

[4] Donoghue JP. Bridging the brain to the world: a perspective on neural interface systems. *Neuron*, 2008, 60(3): 511 - 521

[5] Kim SP, Simeral JD, Hochberg LR, *et al*. Neural control of computer cursor velocity by decoding motor cortical spiking activity in humans with tetraplegia. *J Neural Eng*, 2008, 5(4): 455 - 476

[6] Borton DA, Song YK, Patterson WR, *et al*. Wireless, high-bandwidth recordings from non-human primate motor cortex using a scalable 16-Ch implantable microsystem. *Conf Proc IEEE Eng Med Biol Soc*, 2009: 5531 - 5534

[7] Vargas-Irwin CE, Shakhnarovich G, Yadollahpour P, *et al*. Decoding complete reach and grasp actions from local primary motor

cortex populations. *J Neurosci*, 2010, 30(29): 9659 - 9669

[8] Zhuang J, Truccolo W, Vargas-Irwin C, *et al*. Decoding 3-D reach and grasp kinematics from high—frequency local field potentials in primate primary motor cortex. *IEEE Trans Biomed Eng*, 2010, 57(7): 1774 - 1784

[9] Bansal AK, Vargas-Irwin CE, Truccolo W, *et al*. Relationships among low-frequency local field potentials, spiking activity, and 3-D reach and grasp kinematics in primary motor and ventral premotor cortices. *Journal of Neurophysiology*, 2011

[10] Suner S, Fellows MR, Vargas-Irwin C, *et al*. Reliability of signals from a chronically implanted, silicon-based electrode array in non-human primate primary motor cortex. *IEEE Trans Neural Syst Rehabil Eng*, 2005, 13(4): 524 - 541

[11] Ojakangas CL, Shaikhouni A, Friehs GM, *et al*. Decoding movement intent from human premotor cortex neurons for neural prosthetic applications. *J Clin Neurophysiol*, 2006, 23(6): 577-584

讲座人简介

John P. Donoghue is the Henry Merritt Wriston Professor and Professor of Neuroscience and Engineering, as well as the Director of the Brown Institute for Brain Science at Brown University where he has been a faculty member for more than twenty-five years. His laboratory investigates the function of cerebral cortex and is also engaged in translational research to create neurotechnologies for people with paralysis to regain independence and control. His basic neuroscience research examines the way populations of neurons in the cerebral cortex convert plans, thoughts or ideas into skilled arm movements. This work combines novel implantable multi-neuron sensors developed in his laboratory and mathematical methods to understand how ensembles of neurons represent information and how they compute new information through their interactions. The knowledge and technical advances gained from this fundamental research is being translated into a human brain computer interface technology, called BrainGate. This neurotechnology has the potential to restore useful functions for people with paralysis, providing control of a computer, an assistive robot; it might also allow the brain to be reconnected to paralyzed muscles or to use revolutionary prosthetic limbs for amputees. BrainGate is designed to physically reconnect the brain to the outside world through a tiny "chip" the size of a baby aspirin, that is implanted in the motor area of the cerebral cortex. Pilot trials of BrainGate have already demonstrated the ability for humans with long-standing and severe paralysis to use their own intentions to use send text messages, reach and grasp objects with a robotic arm, and drive a wheel-chair. Moving these early stage demonstrations to real world use is one of next goals of the team of researchers Donoghue leads.

Professor Donoghue was the founding chairman of the Department of Neuroscience in the Alpert Medical School at Brown, a position he held for thirteen years. The Brown Institute for Brain Science, which he founded and leads, unites more than one hundred Brown faculty members in interdisciplinary basic, translational and clinical research on the nervous system. Dr. Donoghue has published widely in the fields of Neuroscience and Neurotechnology along with his many undergraduate, graduate and postdoctoral students who are now well-recognized, independent scientists and clinicians. He was also a co-founder and scientific director of Cyberkinetics, a start-up company (now no longer operating) that played a key role in translating BrainGate to human pilot clinical trials. Dr. Donoghue' research has been honored by Germany's Zülch Prize, the "In Praise of Medicine" award from Erasmus University, the 2010 Pioneer in Medicine award from the Brain Mapping Society, and a Jacob Javits award from the NIH. In addition, the BrainGate project won the 2004 Innovations Award for Neuroscience from Discover Magazine, and the 2007 R&D 100 Award, among many other awards. Donoghue's work has been featured widely in the media, including the New York Times, the CBS television show 60 Minutes, PBS Frontline, Discovery and History channel documentaries and other international media. He has served on numerous United States federal panels for the National Institutes of Health, the National Science Foundation and the Department of Defense. He is also a member of the National Research Advisory Council of the Department of Veterans Affairs, and a Fellow of both the American Institute for Medical and Biomedical Engineering and the American Association for the Advancement of Science.

心-脑科学研究前沿——神经工程

Hideaki KOIZUMI(小泉英明)

(Fellow, Hitachi, Ltd. Member, the Engineering Academy
of Japan Member, Science Council of Japan)

摘要：无损脑功能成像技术使得通过实证主义对人类复杂的脑功能的研究成为可能，并逐渐揭露关于人的意识的各种问题。对复杂的人脑功能的研究已经在心理学、精神医学、行为学、哲学和语言学等多个领域展开。无损脑功能成像技术的出现大大缩短了以心智为主题的学科和神经科学之间的距离。一些以前很难被划入神经科学定义范围之内的相关学科，现在可以被列入脑科学中。跨学科的脑科学研究方法希望能够建立一个连接与融合脑科学（自然科学）、社会科学、人类学和人文学的新学科。这个新学科将包括"脑科学与教育"、"脑科学与伦理学"、"脑科学与人文学"、"脑科学与经济学"、"脑科学与社会规范"和"脑科学与安全"。回顾科学史，望远镜的发明大大促进了天文学的发展，生物学的进步则因显微镜的发明而加快。实验的测量和观察是科学发展的驱动力。与此相似，"脑显微镜"技术的发展被认为是进一步推动脑科学发展的关键因素。迄今为止，传统神经科学和脑功能测量科学一般局限于研究个体的脑功能。尽管样本容量可能增大，但是之前的研究未曾涉及一组被试的脑与现象之间的功能交互性。现在这种模式正在发生改变。近红外光脑地形图系统是半导体装置，能够做得很紧凑，一种可穿戴式光学成像系统原型已经开发出来，不久就能投入使用。这项技术能够几乎实时地观察多个大脑的功能及其相互作用，对研究社会科学中不同科学组群之间的桥接和联合关系将起到不可或缺的作用。

关键词：神经工程；近红外光谱；脑功能成像

A Frontier of Mind-Brain Science: Neuro-Engineering

Hideaki KOIZUMI

(Fellow, Hitachi, Ltd. Member, the Engineering Academy
of Japan Member, Science Council of Japan)

Abstract: Recent progress in noninvasive brain function imaging technique, which allows empirical studies to be conducted on the sophisticated human brain functions, is gradually uncovering the problem of willingness. Research on human's original sophisticated brain functions has been studied from the standpoint of psychology, psychiatric medicine, praxeology, philosophy and linguistics. In particular, the distance between the disciplines which handle mind issues and neuroscience has narrowed rapidly due to the appearance of the noninvasive brain function imaging technique. Current Brain-Science is an attempt to create new academic disciplines by bridging and fusing brain-science (natural science) with the social science, the humanities and the arts. Such new disciplines include "Brain-Science & Education", "Brain-Science & Ethics", "Brain-Science & the Arts" and etc. Looking back at the history of science, empirical measurements and observations are truly the driving force of scientific development. In the trans-disciplinary Brain-Science study, further development of non-invasive brain function imaging techniques, like "brain-scope", is a key factor. Today, a paradigm shift is taking place. The aforementioned NIR – OT system can be made more compact because, in principle, it is a semiconductor device, and a prototype wearable optical topography (WOT) system has already been developed. Hopefully, it will be in practice use soon. This technological achievement should allow the observation of the functions and interaction of numerous brains at the same time-almost on a real-time basis. The author believes that this technique will become indispensable for bridging and fusing society as agglomerations of groups with the social sciences.

Keywords: Neuro-engineering, Prototype wearable optical topography, Brain function imaging

1　引言

在还原论的基础上进行学科间的整合(Integration Over Reductionism)在 21 世纪具有非常重要的意义。通过动态地增加新的学科来创建新的研究领域,通过科学家、学者和从业人员的相互合作,在脑科学和分析科学的基础上,我们开展了新的科学问题的研究:以人为中心的安全问题和人的幸福的研究。该研究方向从计算神经科学、实验神经科学、临床神经病学、电子工程学和活神经组织的信号处理等领域汲取养分,并包含了机器人学、计算机工程学、组织工程学、材料科学和纳米技术等学科中的一些内容。这个学科的一个基本研究内容是脑信号的测量。脑信号的测量技术可以分为有创(如植入电极)或无创(如头皮脑电、近红外光谱、核磁共振)两种。我们的研究团队在无损脑信号测量方面做了大量的工作,研制了日立公司的第一台核磁共振仪,发明了近红外光谱成像仪。近红外光谱仪具有便携式、安全、价格低和无创等特点,能应用于多种场合下大脑活动的研究。我们将这些技术应用在婴幼儿群体脑功能治疗和研究、社会协作和决策研究、教育工程、脑-机接口等领域,推动了神经工程技术在这些领域的发展。表 1 列出了过去 40 多年来在人本安全方面进行的相关研究。

表 1　过去 40 多年来在人本安全方面进行的相关研究

Original achievements	Approximate time
Directing cohort studies based on Brain-Science Japan's children's study and various cohort studies	2005—2010
400th Anniversary of the foundation of PAS pontifical academy of science	2005—2010
Propose concept:"Brain-Science & Education" improvement of general & special education	1990—2000
Creation of optical topography noninvasive, low restriction brain function imaging	1990—2000
Patient-friendly MRI diagnosis system grand prize of good design prize(Health & Welfare)	1985—1995
Visualization of blood vessels using MRI without any contrast agent (MR Angiography) preventive medical checkup of the brain	1985—1990
Sonic-spray mass spectroscopy ultra-high sensitive moleculeanalysis in human & environment	1995—2000
Polarized zeeman atomic absorption spectrometry detection of trace elements in human & environment	1975—1985

2　环境安全监测仪器

在环境安全监测方面,我们在 20 世纪 80 年代开发了塞曼效应汞分析仪 (图 1)和塞曼效应原子吸收分光光度计。塞曼效应汞分析仪的原理是基于 Zeeman 效应原子吸收光谱,将汞灯放置在一个永久磁场中,发射共振发射波, 其共振线在磁场的作用下分裂成三个极化的 Zeeman 组分。这些组分与汞蒸 气的浓度有特定的关系,并且不受水蒸气、灰尘或是其他有机、无机气体的干 扰,通过校正能够准确地测量空气中汞的浓度。塞曼效应原子吸收分光光度 计是利用塞曼效应进行背景校正的原子吸收光谱仪,分为光源调制型和吸收 源调制型两大类型。它与单光束原子吸收光谱仪的构造相似,只是在光源或 原子化器上施加 $10 \times 10^3 \sim 15 \times 10^3$ 高斯[1 高斯(G)$= 10^{-4}$特斯拉(T)]的强 磁场,使发射线或吸收线发生分裂。分裂后波长不变的 π 线作为分析线,测定 总吸收(原子吸收＋背景吸收);波长改变的 ±σ 线作为参比线(为背景吸收)。 它有良好的抗干扰能力,而且可在 190～860nm 全波段内使用。这些仪器在 空气质量检测、环境安全监测方面应用广泛。

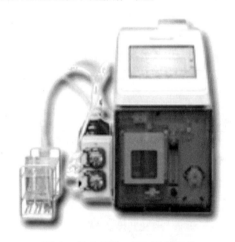

图 1　日立 HG‑400 汞分析仪

3　无损脑功能成像技术

近年来,随着科学技术的迅速发展,认知神经科学领域涌现了一批功能强大的脑成像技术,如功能性磁共振技术(fMRI)、近红外光谱成像技术(NIRS)、正电子发射层析技术(PET)、单光子发射计算层析技术(SPECT)、头皮脑电(EEG)和脑磁图(MEG)等。这些技术大大拓展了认知神经科学领域的研究。不同的测量技术具有不同的时空分辨率。我们在功能性磁共振技术和近红外光谱技术方面做了大量研究,在 1983 年研制了日立公司的第一台核磁共振仪,1985 年研制出血管造影仪,1989 年研制了 3D 核磁共振仪,1992年研制了增强型功能性核磁共振仪,并在同年首次获得运动想象时运动区的功能性核磁共振数据。除核磁共振技术外,在 1991 年研发了近红外光谱成像仪,具有便携式、安全、价格低和无创等优点,能在多种场合下研究大脑皮层的活动情况,可实现多方面的应用,包括工作绩效评价、麻醉状态监测、神经康复、脑-机接口、心理健康治疗等。

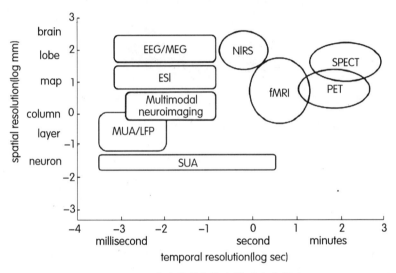

图 2　不同脑功能成像技术的时空分辨率

3.1　磁共振成像技术原理

核磁共振成像(MRI)是利用原子核在磁场内共振所产生的信号经重建成像的一种技术。人体组织含有大量的水和碳氢化合物,氢核的核磁共振灵敏度高、信号强,成为人体成像元素的首选。核磁共振信号强度与样品中氢核密度有关,人体中各种组织间含水比例不同,即含氢核数的多少不同,因此信号强度有差异,从而把各种组织分开,这就是氢核密度的核磁共振成像原理。人体不同组织之间、正常组织与该组织中的病变组织之间氢核密度、弛豫时间 T1 和 T2 这三个参数的差异,是 MRI 用于临床诊断最主要的物理学基础。

当施加射频脉冲信号时,氢核能态发生变化,射频过后,氢核返回初始能态,共振产生的电磁波便发射出来。原子核振动的微小差别可以被精确地检测到,经过进一步的计算机处理,即可能获得反映组织化学结构组成的三维图像,从中我们可以获得包括组织中水分差异以及水分子运动的信息。这样病理变化就能被记录下来。人体内器官和组织中的水分并不相同,很多疾病的病理过程会导致水分形态的变化,即可由磁共振图像反映出来。

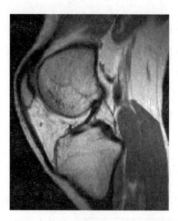

图3　日立开放式 MRI 最先实现的 2048×2048 的图像重建矩阵

MRI 所获得的图像非常清晰精细,大大提高了医生的诊断效率,避免了剖胸或剖腹探查诊断的手术。由于 MRI 不使用对人体有害的 X 射线和易引起过敏反应的造影剂,因此对人体没有损害。MRI 可对人体各部位多角度、多平面成像,其分辨力高,能更客观、更具体地显示人体内的解剖组织及相邻关系,对病灶能更好地进行定位定性,对全身各系统疾病的诊断,尤其是早期肿瘤的诊断有很大的价值。

3.2　近红外光谱技术

近红外光谱技术以氧合血红蛋白、脱氧血红蛋白和细胞色素氧化酶等为指标,考察与神经元活动、细胞能量代谢以及血流动力学相关的大脑功能,所以也称近红外功能成像技术(NIRS)。这一技术具有时空分辨率较高、便携性强、价格低廉和无创等优点,在认知神经科学和医学等的研究中得到了越来越广泛的应用。

近红外光谱技术检测脑功能的主要神经生理学和神经能量学基础是神经-血管耦联机制,即大脑的血流供应会随着功能活动的局部变化而进行局部响应。当大脑处于激活状态时,会引起局部脑血流与氧代谢率改变,从而引起相应区域内血氧浓度的变化,因此通过测量组织血氧状态,即可间接评价大脑功能活动。

NIRS 检测组织血氧基于以下原理：第一，人体组织对近红外光 700～900nm 呈现高度前向散射和低吸收特性，使光子能够穿透几个厘米深度；第二，含氧血红蛋白和去氧血红蛋白在近红外光谱区的吸收系数有明显差异。利用该谱区两个波长的光检测组织光吸收变化，就可以分别计算出血液中含氧血红蛋白和去氧血红蛋白浓度以及血容量的变化。NIRS 成像示意图如图 4 所示。

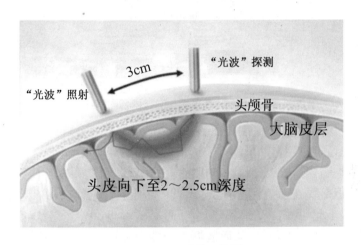

图 4　NIRS 成像示意图

4　应用

肌萎缩性侧索硬化症（Amyotrophic Lateral Sclerosis，ALS）是一种运动神经和肌肉退化的疾病。完全瘫痪的病人不能用眼睛、嘴巴和手等与外界通信，因为这些器官依赖于肌肉系统，但是听觉系统不依赖于肌肉，所以病人可以听到声音（图 5）。长期瘫痪的 ALS 病人是否还有意识，如果有意识那他们在想些什么，是否存在和他们沟通的方法？针对这些问题，我们与 Tokai 医学院开展合作，采用光成像技术进行了大量研究。在病人试图说话的时候，布洛卡区（与发音有关的脑区）有明显的神经活动；在病人听别人说话的时候，在韦尼克氏区（与语言理解有关的区域）有相关的神经活动。通过在其他的功能区进行类似的大量实验，结果证明，长期完全封闭的 ALS 病人仍然存在意识。此后我们又研究了与这些病人进行沟通的方法。病人被告知，如果他对一个问题的回答是肯定的，则想象右手的抓握运动，如果回答是否定的，则不进行想象。通过分析运动区不同的光学图像模式，能破解病人想回答“是”或“否”

的意愿,从而帮助他们与外界进行简单的交流。

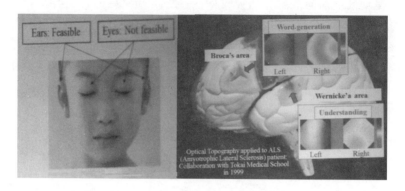

图5　肌萎缩性侧索硬化症病人示意图(左)和语言区近红外光谱成像图(右)

在光成像技术的基础上,我们在2006年开发了一个脑-机接口的演示系统。该系统通过对用户的近红外光谱信号进行解码,将解码结果传给模拟控制器,最终控制模型火车。这种脑机接口技术在教育和康复方面具有很大的应用前景,但是要在现实世界中对这些问题进行实际的研究,需要便携式的测量仪器。鉴于此,我们在2007年研制了穿戴式的近红外光谱仪,能同时记录24个不同被试大脑的信号,从而能够进行社会关系中大脑活动的协同研究。我们对教学活动中老师和学生的大脑信号非常感兴趣,因此将这套系统用于研究师生互动和提高教育质量的问题上,我们认为这正是神经工程的研究领域。

图6　基于近红外光谱成像的脑-机接口演示系统(左)和近红外光谱协同记录系统(右)

近红外光谱成像技术非常安全,能用于采集新生婴儿的大脑信号,并应用于新生婴儿大脑功能重建的研究。结果发现新生婴儿的大脑具有令人难以置信的可塑性:在大脑内囊几乎完全丧失的情况下仍然能够运动,在左脑严重损伤的情况下仍然能够说话,在丧失小脑的情况下仍然能够进行复杂运动。这个研究结果2000年在BBC的"Tomorrow's World"栏目中广播过。

关于新生婴儿的另一项研究是母语和大脑活动的关系。我们与巴黎国家认知实验室(National Lab. For Cognitive Science)的 J. Mehler 小组进行合作,在征得法国国家伦理委员会的同意后对新生婴儿进行了实验。由于在法国进行研究,所以母语采用法语。我们发现,即使是刚出生不到 5 天的新生儿,在他们听到母语的时候,左脑会出现比右脑更加明显的活动。随后我们非常谨慎地重复了这种实验,采用意大利语作为母语,取得类似的结果,并且在 *PNAS* 上发表了论文。

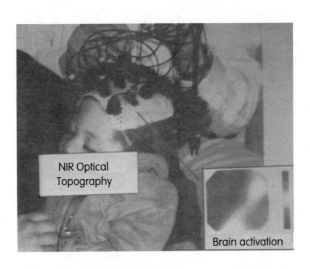

图7　近红外光谱成像用于婴幼儿脑功能重建的研究

另外,我们于 2008 年在 *Neuron* 上发表了一篇关于社会奖励的文章。研究人员设计了一个实验范式,告知被试其他被试将通过他的一段录像去判断他这个人的可信度,因此,被试会为了试图得到别人的好评而尽量好好表现,同时其大脑的尾状核区域也会处于活跃状态,该结果可于将 fMRI 图像处理后得到。这篇文章通过功能性磁共振成像的手段,研究了社会奖励(social reward)和金钱奖励(monetary reward)对人脑纹状体区域(striatum)的影响,证明了社会奖励和金钱奖励同等重要,甚至更重要一些。其中社会奖励指的是得到别人对自己的认同,而且我们相信这正是教育学的本源。

同时,为了方便数据的采集和实验的进行,我们的团队专门做了一种可移动的实验用卡车(图 8),车上装备有多台 64 通道的近红外光谱仪(Near infrared optical topography,NIR – OT)以及 64 通道的 EEG(头皮脑电)信号采集仪器。我们把这个可移动的实验室开到了很多学校,在那儿待上一个星期左右的时间,通过这种方式,科研人员测试了很多学生,主要进行了第二语言学习方面的研究,并把相关的研究结果发表在了期刊 *Cerebral Cortex* 上。在日本,第二语言学习实际上指的就是英语的学习。毫无疑问,如何提高学习英语的效率是一个很重要的教育议题,因此我们认为这种研究有较大的应用价值。

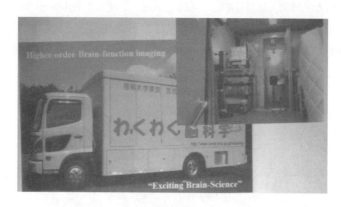

图 8　可移动的脑科学实验卡车

我们的团队在日本有很多的研究项目,主要分为基础理论研究和应用研究两个大方向。在基础理论研究方面,主要有"学习科学与脑科学研究"和"基于脑功能发育的教育学研究"两大项目,这两个项目中我是作为项目的顾问。在应用研究方面,主要有"脑科学和教育学(类型 1)"与"心理和脑的健康发展"两个大的项目,在这两个项目中,我都是项目的负责人。其中"脑科学和教育学(类型 1)"的研究是从 2001 年开始的,这是所有项目中第一个正式项目,也是第一个脑科学与教育的交叉研究项目。

下面是我们所在的日本科学技术研究所在"脑科学与教育"方面的研究发展历程:从 1996 年的一个国际论坛逐渐开始将脑科学与教育相结合的研究工作,于 2001 年到 2003 年开展了"脑科学与教育"的第一阶段的研究,这一阶段的研究是一种自顶向下的研究方式,共包括 12 个相关的项目。2003 年开始了"脑科学与教育"第二阶段的研究工作,这一阶段的研究主要是以自底向上的群体性研究为主,共有 6 个相关的项目。随后又开展了日本儿童学习问题的研究,该研究是一种自顶向上的群体性研究,其中包括一个大型的项目。现在这种将脑科学与教育学结合起来的新想法已经纳入了日本的国家基础计划。

5　总结

在过去 40 多年的研究过程中,我们在环境、医疗、大脑测量领域开展了前沿性的研究,代表性的工作包括偏振原子吸收光谱测定法(PZAA)、医疗测量仪器如磁共振血管造影法(MRA)和功能磁共振成像术(fMRI)以及近红外光谱图技术(NIRS-OT)。其中 NIRS-OT 技术具有便携式、安全、价格低和无创的特点,能适用于各种不同的场合。同时,我们也致力于将这些技术应用在婴幼儿群体脑功能治疗和研究、社会协作和决策研究、教育工程、脑-机接口等领域,推动了神经工程技术在这些领域的发展。2003 年的 MIT 技术综述评价了我们在脑科学方面进行的研究项目,认为这些研究在通过脑成像技术提高教育质量方面打破了传统模式,并将其评选为 2003 年的四大突破性研究之一。

　　回顾科学史,由于望远镜的发明大大促进了天文学的发展,生物学的进步则因显微镜的发明而加快。实验的测量和观察是科学发展的驱动力。与此相似,"脑显微镜"的发展被认为是进一步推动脑科学跨学科研究的关键因素(图9)。继哈勃望远镜、电子显微镜之后,"脑显微镜"的时代已经来临。我们开发的可穿戴式近红外光谱成像仪具有便携式、安全、价格低和无创的特点,广泛应用在不同领域不同场合的研究,对脑科学的研究起到了一定的推动作用。

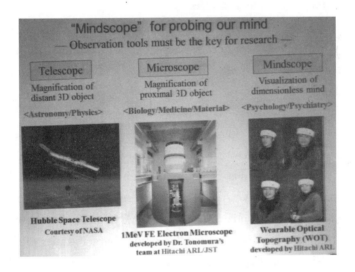

图 9　"脑显微镜"帮助探测人的意识

参考文献

[1] What is Neural Engineering? http://iopscience. iop. org/1741 - 2552/4/4/E01

[2] He B, Yang L, Wilke C, and Yuan H. Electrophysiological imaging of brain activity and connectivity-challenges and opportunities. *IEEE Transactions on Bio-medical Engineering*, 2011, 58: 1918 - 1931

[3] Izuma K, Satio DN, Sadato DN, Sadato N. Activation of striatum

by monetary and social network. *Neuron*, 2008

[4] Koizumi H, Yamashita Y, Maki A, *et al*. Higher-order brain function a-
nalysis by trans-cranial dynamic near-infrared spectroscopy imaging.
Journal of Biomedical Optics, 1999, 4:403

[5] Pea M, Maki A, Kovac D, *et al*. Sounds and silence: an optical topography study of language recognition at birth. Proceedings of
the National Academy of Sciences of the United States of America,
2003, 100(20): 11702

[6] Utsugi K, Obata A, Sato H, *et al*. Development of an optical brain-
machine interface, 2007:5338 – 5341

[7] Maki A, Yamashita Y, Ito Y, *et al*. Spatial and temporal analysis
of human motor activity using noninvasive NIR topography. *Medical Physics*, 1995, 22:1997

[8] Taga G, Asakawa K, Maki A, *et al*. Brain imaging in awake infants
by near-infrared optical topography. Proceedings of the National
Academy of Sciences of the United States of America, 2003, 100
(19): 10722

[9] Watanabe E, Maki A, Kawaguchi F, *et al*. Non-invasive assessment of language dominance with near-infrared spectroscopic mapping. *Neuroscience letters*, 1998, 256(1): 49 – 52

[10] Koizumi H, Yasuda K. New Zeeman method for atomic absorption
spectrophotometry. *Analytical Chemistry*, 1975, 47(9): 1679 – 1682

[11] Koizumi H, Yasuda K, Katayama M. Atomic absorption spectrophotometry based on the polarization characteristics of the Zeeman
effect. *Analytical Chemistry*, 1977, 49(8): 1106 – 1112

讲座人简介

Hideaki KOIZUMI is currently a Fellow
of Hitachi, Ltd. He also serves as the Director of the Research & Development Division
"Brain-Science & Society" at the Research Institute of Science and Technology for Society
of the Japan Science and Technology Agency,
and a Director of the Institute for Seizon and
LifeSciences. He is a Vice President of Japa-

nese Society of Baby Science, Fellow of the Chemical Society of Japan, an Academician of the Engineering Academy of Japan, anda member of the Science Council of Japan. Other public positions include Specialist Committees of the Central Council for Education and the Atomic Energy Commission. Internationally, he is a founding board member of the International Mind, Brain and Education Society. He was the 55th President of the Japan Society for Analytical Chemistry, a Director of the Japan Neuroscience Society, a Director of the Chemical Society of Japan, the Auditor of National Institute for Environmental Studies, an Adviser of the OECD/CERI initiative "Learning Sciences and Brain Research", and also served as a Visiting Professor at various faculties and graduate schools in the University of Tokyo and others.

Upon graduating from the University of Tokyo in 1971, Koizumi joined Hitachi. In 1976, he received a doctoral degree in physics from the University of Tokyo for hiswork on the creation of the Polarized Zeeman Atomic Absorption (PZAA) spectrometry, which was also nominated as one of the 50 most significant patents at the centennial of the foundation of the Japanese patent system in 1985. This technology was commercialized by Hitachi, and about 10,000 systems based on this principle have been shipped to 25 countries since 1976. As a Guest Worker at the National Bureau of Standards of the U. S. Department of Commerce, he worked on the certification of standard reference materials (SRMs) using his PZAA method in 1976, and as a Guest Research Physicist (faculty) at the Lawrence Berkeley Laboratory of the University of California in 1977 - 1978. After returning to Japan, he was appointed leader of Hitachi's MRI (Magnetic Resonance Imaging) development project. He and his colleagues also developed MRA (Magnetic Resonance Angiography) in 1985, fMRI (functional Magnetic Resonance Imaging) in 1992, and NIRS - OT (Near-Infrared-Spectroscopy Optical Topography) in 1995. He has received many prizes and awards such as the Ohkohchi Prize(three times), the Science and Technology Minister's Prize (twice), and the R&D - 100 Prize (twice). He has applied for more than 400 patents.

运动学习和功能康复中脑皮层 神经活动的适应性

何际平

(Harrington Department of Bioengineering, Department of Electrical Engineering, Ira A. Fulton School of Engineering, Arizona State University)

(国家千人计划特聘教授,神经接口与康复技术研究中心,华中科技大学, 武汉,中国)

摘要:使用神经接口,我们可在数月甚至数年时间内同时监控大量的独立皮层及其周边神经元的活动。同时,这一技术也为运动学习任务或者为外伤或神经性疾病引起的运动功能康复治疗过程中对中枢神经系统内的动态机制的探索提供了可能。对神经损伤疾病(如脑卒中、脊髓损伤、帕金森病等)的治疗过程中大脑所展现的自适应机制进行研究,可能会促使发明更具创新性和高效性的疗法。在本文所描述的两个例子中,均使用了神经接口进行监测,并研究了运动学习过程中神经元的可塑性及脊髓受损后帮助重建行走功能所进行的康复训练过程。通过这些实验,我们可以推断:非人灵长类动物模型的实验结果可以用于临床功能相关技术,而使用功能性相关的神经可塑性则可提高对中枢神经系统疾病的疗效。

关键词:神经接口技术;医疗康复;脑皮层适应性;脊髓损伤(SCI)

Adaptation in Cortical Neuron Activity during MotorLearning and Functional Rehabilitation

Jiping He

(Harrington Department of Bioengineering, Department of Electrical Engineering, Ira A. Fulton School of Engineering, Arizona State University)

(Ctr Neural Interface/Rehabilitation Technologies, Huazhong University of Science and Technology, Wuhan, China)

Abstract：Neural interface technologies have advanced to allow simultaneous monitoring of large population of individual cortical and peripheral neurons for months or even years. This technique presents opportunity to investigate dynamic processes in the central nervous system during learning of new motor behavioral tasks or rehabilitation for motor function recovery after trauma or neurological diseases. To understand the adaptation process during treatment of neural trauma of diseases，such as stroke，spinal cord injury or Parkinson's disease，etc.，may lead to discovery of innovative and more effective therapeutic approaches. This presentation will describe two examples of using the neural interface to monitor and learn of neural plasticity during motor learning and motor task training after spinal cord injury for recovery of walking. Though the experiments were conducted in non-human primate models the findings may lead to clinical techniques to induce functionally relevant neural plasticity in the central nervous system for effective therapy.

Keywords：Neural interface technology，Medical recovery，Cortex adaptation，Spind cord injury(SCI)

1 引言

近年来，随着科学理论与技术设备的迅速发展，利用植入式电极直接获取脑电信号并进行分析已逐渐成为了神经编解码领域的主流研究方法。绝大部分的研究工作集中在运动控制方面，而事实上，运动控制系统的运作模式异常复杂：现有研究成果已经证明，实现运动控制过程需要大脑中大部分区域的相互协作，不能把对运动控制的研究和感知信息分割开来。

除了这类学术研究以外，建立一个完整的感知、认知与运动控制的综合系统还可以为脊髓损伤(SCI)患者的医疗康复提供额外的帮助。借助该综合系统，医生可为每个 SCI 病人设计特定的康复疗程与治疗方法。

2 感知、认知、运动控制的综合系统

本领域的研究人员所采用的通用方法是建立在运动感知信息结构基础之上，如图 1 中所示，我们将电极植入猴子大脑中相关的运动皮层区域(包含前运动区域和感知区域),然后训练猴子做简单的伸手、抓取动作,以及控制机械完成交互。该实验范式可概括为从猴子的大脑中采集信号,记录相关的肌电

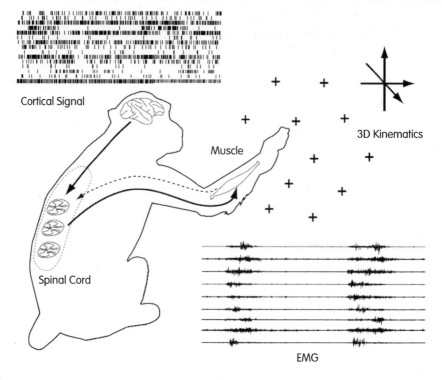

图 1 运动皮层与伸展运动的链接通路

(EMG)信号(包括手肌和腿肌),同时在虚拟环境或物理环境中训练猴子做三维的伸手和抓取动作,观察并分析不同条件下信号的区别,如图 2 所示。

实验中所做的是训练猴子在特定的任务中做出特定的反应行为,然后训练它们控制真实或者虚拟的物体。从猴子大脑皮层中采集得到的信息在经过通用方法分析后可从中分离出运动与感知的相关性,建立对应的方程。

在完成这一阶段的工作后,要使用建好的方程匹配脑电信号驱动机械手完成伸展、抓取动作,首先需要解决的问题是确认控制的精度并了解猴脑的适应能力,这两个问题决定了该如何设计相应的训练方案。

在实验的开始阶段,我们在训练猴子的同时固定住它的手臂。猴子要做到的是像控制自己的手一样控制球,在这个阶段仍然可以获取相关的肌电信号。在经过一段时间的训练后,我们发现猴子在控制球移动时不再发出EMG 信号。该实验证明,大脑本身存在一种转换机制,可以使原来用于驱动肌肉的大脑信号不再用于驱动相应的肌肉运动,转而成为虚拟环境中的控制信息。

事实上，在实际的研究实验中干扰信号与错误信号的出现是不可避免的。改变控制杆的运动模式，或者提高物体移动的随机性，都将促使猴子必须在实

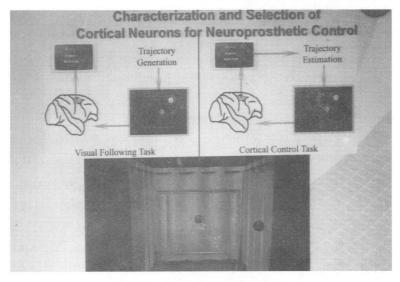

图 2　虚拟场景中的三维交互

验中做出正确的预测。实验证明，猴子只需要经过几次的尝试即可越过干扰信号，直接伸手到正确的目标位置完成实验。图 3 展示了猴子在控制杆以不同的速度改变位置的情况下展现出来的适应能力。

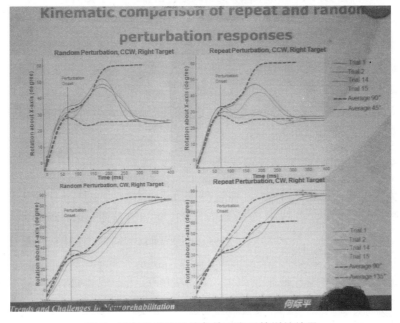

图 3　重复及随机干扰条件下猴子的训练结果

左边的图是没有加入干扰的结果。右上方的图所代表的实验中加入了随机的干扰,右下方加入的是重复的干扰。猴子的学习过程很快:只需要两次尝试,它们就能知道它们需要完成什么动作。图3表明了加入干扰后猴子的神经元活动与原先有所区别,采集自皮层神经元的信号印证了这些差异的存在。在加入重复干扰的实验中,猴子完成数次实验动作后脑皮层的相关神经元活动非常活跃,再次的实验中我们可以观察到在猴子的动作初期,相关脑皮层中的神经元峰电位已经提早达到兴奋的程度。这表明经过几次重复干扰后的猴子具有了预测的能力,预先判断出了杆子的最终位置。

同时我们还完成了另外一组实验(让猴子做伸展运动的时候直接干预它的手部动作)用于对比。对猴子运动皮层区域 M1 中的神经元活动状态进行观察,我们可以发现,在人为干扰出现前,在猴子脑区的对应神经元上已经出现了一定程度的兴奋。这表明 M1 区域中神经元的兴奋状态是可以进行预测的,并且不同于简单的感知信号,当插入干扰时,大脑会在感知皮层中产生很强的神经兴奋作为回应,而随着干扰的剧烈程度上升,感知灵敏度开始下降,感知能力遭到削弱。由此,我们可以得出如下结论,由于作用域的不同,运动皮层和感知皮层对抗干扰有着不同的适应策略。该结论对神经信号编解码工作有一定的启发作用。

3 脊髓损伤对脑部活动及脑电信号的影响

下一步研究主要观察脊髓损伤如何影响脑部活动及脑电信号,实验设计的关键是训练猴子在不同的外界条件下(如速度、外力)完成康复,随后完成相关的研究工作。实验设计的目的是试图找出是否有特殊的康复治疗和训练方法可以帮助提高脊髓损伤的功能性修复,及其对脑部活动和脑电信号的影响。

实验过程及内容如下:

3.1 脊髓分段及/或修复手术

为猴子实行椎板切除手术:在脊髓的 T8 段右半边制造半截约 4~5mm 的缝隙,并清除在这条狭缝内的所有组织。同时,从猴子的左腿中取出约 15cm 的腓肠神经,将其切割成 5~6mm 长度的小片,填充入受损的脊髓两端,用以提供结构性支撑并尽可能减小由于脊髓损伤带来的伤害。此外,还有选择地填入含有酸性成纤维细胞生长因子的纤维蛋白混合物,以帮助稳定末梢神经的嫁接并修补受损的神经网络。

3.2　采集的实验相关数据

数据包括肌电信号、脑皮层活跃程度、视频及与神经元光栅同步后的视频。

3.3　猴子的康复结果

猴子在手术后两周便开始了康复训练。训练从普通的腿部活动开始,包括伸展及踏板运动。由于猴子的小腿肌肉非常平滑,因此可以以任意方式进行移动。在经过纯化训练过程后,可以观察到一开始猴子的状态并没有明显的变化,但是6周后,猴子开始试图行走,这可以视作功能上的逐渐康复。10周后,它们基本上可以正常行走,虽然在运动学上有所残缺,但还是可以视作达到了基本康复。

3.4　功能康复与皮层控制

如图4所示,图左边是猴子在脊髓受损前的肌电信号图,中间记录的自受损6周后康复阶段的肌电信号,最后则是14周后恢复阶段的肌电信号。

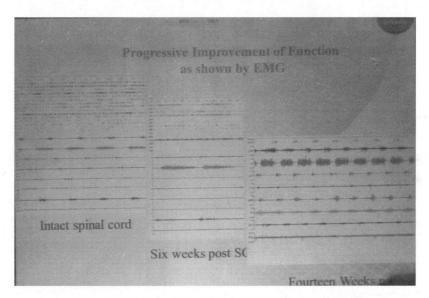

图4　SCI手术前、术后6周及术后14周的EMG信号图

在脊髓受损前,脑皮层的各个区域独立运作,但是在脊髓受损后,腿部相关区域的皮层神经元运作模式趋向同步。训练可明显促进功能康复,使得疗程的质量也得到了显著的改善。由于受到所采用的(植入式)技术的限制,手术后至6到7周时间内所采集得到的信号质量是最高的,4个月内的信号也可以勉强使用,但是24周,也即6个月后,我们就无法采集到任何信号了。

从康复过程中我们可以发现,手术8周后猴子大脑皮层的神经元活动得到了同步,神经元间构建了统一的工作环路,换句话说,猴脑已经产生了自适应现象,以促成功能康复并重建对运动区域的控制(图5)。

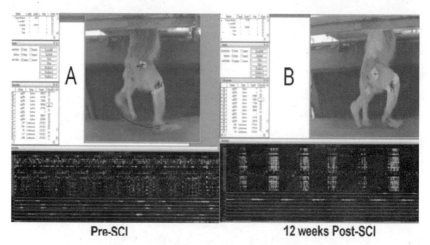

图5　SCI手术前、术后12周的康复结果

4　总结

实验证明,猴子的大脑皮层神经元的活跃性在经过受损、康复训练后均发生了变化,一方面,由于功能上存在差异,大脑中运动与感知的神经元集群有不同的自适应能力。运动区域作用于预测与控制,而感知区域在我们的实验中随着干扰的增加敏感性有所降低;而在另一些实验中情况则正好相反。另一方面,在脊髓损伤发生后,经过康复训练的猴子大脑自发地进行了一些功能性相关的自适应调整。我们的课题将来所面对的问题是:在临床治疗/医疗康复中,如何更好地促进或抑制条件变化或大脑可塑性以取得更好的功能康复效果?

参考文献

[1] Balasubramanian S, *et al*. RUPERT：An exoskeleton robot for assisting rehabilitation of arm functions. in：*Virtual Rehabilitation*, *2008*：163 - 167

[2] Song D, *et al*. Predicting EMG with generalized Volterra kernel model. in：*Engineering in Medicine and Biology Society*，*2008*.

EMBS 2008. 30th Annual International Conference of the IEEE, pp. 201－204

[3] Duff M, *et al*. An Adaptive Mixed Reality Training System for Stroke Rehabilitation. *Neural Systems and Rehabilitation Engineering*, *IEEE Transactions on*, 2010, 18: 531－541

[4] Ma C, He J. A method for investigating cortical control of stand and squat in conscious behavioral monkeys. *J Neurosci Meth*, 2010, 192: 1－6

讲座人简介

Dr. Jiping He born in Shanghai, China. B. S. degree in Control Engineering from Huazhong University of Science and Technology, Wuhan China, in 1982, M. S. and Ph. D. degrees in Electrical Engineering from University of Maryland, College Park, 1984 and 1988. Post-doctoral fellow in Center for Biological Information Processing and Artificial Intelligence Laboratory, Massachusetts Institute of Technology, Cambridge, MA.

In 1990, He joined Functional Neurosurgery Division of Thomas Jefferson University (TJU), Philadelphia, PA, as a research assistant professor, and also an Adjunct professor of Physical Therapy. He was a visiting scientist at Human Information Processing Program of Princeton University, Princeton, NJ during 1991 and 1992. He has been a faculty at Arizona State University, Tempe, AZ since 1994 and is now Professor of Bioengineering, Director of Center for Neural Interfaces Design. He also held visiting and honorary professorship in several universities.

Dr. He's primary research interests include the application of advanced control theory to the analysis and control of neuromuscular systems for posture and movement, implantable neural interface technology, cortical and spinal cord recording and stimulation for sensorimotor adaptation and control with application to advanced and smart prosthetics, application of robotics and virtual reality research to neural rehabilitation and prosthetic devices.

浙江大学求是高等研究院
脑机接口研究进展

郑筱祥

（浙江大学求是高等研究院，杭州，310027，中国）

摘要：脑机接口作为神经工程研究的前沿领域，旨在脑与外部设备之间建立一种新型的信息交流与控制通道，实现脑与外界的直接交互，它的发展对于脑与认知、智能信息处理、仿脑工程和人工智能等有重要的科学意义。浙江大学求是高等研究院成立后的四年多来，研究团队紧紧围绕脑机接口中信息"解析"与"交互"两大核心科学问题，充分发挥多学科交叉研究的优势，分别建立了基于啮齿类动物（大鼠）和非人灵长类动物（猕猴）的植入式脑机接口研究平台，在"复杂环境下的大鼠导航"、"小动物双向脑机接口系统"、"基于大鼠运动神经解码的神经控制"和"实时运动神经解码理论与方法"等项目的研究中取得了重要的突破，在脑机接口研究方面实现了快速而稳定的发展。

关键词：双向脑机接口；动物机器人；初级运动皮层；神经解码

BMI Research at Qiushi Academy for Advanced Studies, Zhejiang University

Xiaoxiang Zheng

(Qiushi Academy for Advanced Studies, Zhejing University, Hangzhou, 310027, China)

Abstract: As a frontier of neural engineering research, Brain machine interfaces (BMIs) aim to established a new pathway for communication and control between brain and external devices and to realize direct interaction. The development of BMIs are highly significant in intelligent information processing in brain, artificial brain engineering and artificial intelligence. Since the establishment of Qiushi Academy for Advanced Studies in 2006, our group focus on the two key science problems：decoding and interaction in BMIs. Based on our research flatforms for BMIs in rodents and non-human primates，we have successfully made significant breakthroughs "rat navigation in complicated environment"，"bi-directional BMIs system for small free moving animals"，"mind control using neural signals from motor cortex of rats" and "theories and methods about realtime neural decoding of motor cortex". These studies help us to achieve sustained and rapid development in BMIs researches.

Keywords：Bi-direction brain-machine interface，Biorobot，Primary motor cortex，Neural decoding

1 引言

神经工程主要研究神经系统信息的产生、编码、存储等过程与机理,以及与人类认知相关的计算、控制和行为感知模型,涉及生命科学、信息科学、工程学等多学科的交叉融合,是第三次生物革命的重要内容。作为 21 世纪生物和信息技术发展及应用的前沿热点研究,脑机接口不依赖于常规的脊髓/外周神经肌肉系统,在脑与外部设备之间建立一种新型的信息交流与控制通道,实现了脑与外界的直接交互。脑机接口的研究对脑与认知、智能信息处理、仿脑工程和人工智能等有重要的科学意义,有利于推动新型信息感知、复杂数据处理、模式识别、认知计算和人机交互等技术的研究与发展,在挖掘人类认知潜能、残障人康复、神经疾病治疗,以及航天、国家安全等问题上都具有重要的社会意义和广泛的应用前景,因此引起了国际学术界的极大关注,成为信息科学与神经科学交叉研究领域的前沿热点问题。

自 21 世纪以来,*Nature* 和 *Science* 等报道了一系列植入式脑机接口的重大成果,相关研究促进了人们对神经系统的认识,极大地推动了信息、认知等科学的发展。特别是美国布朗大学 John P. Donoghue 教授领导的团队研发的 BrainGateTM 获得了 FDA 的认证,并先后成功地在 6 名高位瘫痪的患者身上进行了临床实验。该系统可将患者运动皮层神经元电信号通过实时信号处理分析,转换成控制外部设备的指令,患者几乎无需训练就可以用意念移动屏幕上的光标或简单地控制假肢。这项具有开拓性的工作,充分表明了植入式脑机接口在临床应用方面具有巨大的潜力。

2　QAAS 在脑机接口方面的工作

浙江大学求是高等研究院成立于 2006 年 10 月，以材料科学、计算机科学、生物医学工程及临床医学等四大骨干学科为依托，围绕"脑机交互技术"，针对神经工程的基础理论与关键技术，从植入式与非植入式脑机接口两个方面开展多学科交叉研究。

成立四年来，浙江大学求是高等研究院紧紧围绕脑机接口中信息"解析"与"交互"两大核心科学问题，充分发挥多学科交叉研究的优势，在脑机接口研究方面实现了快速而稳定的发展。研究团队分别建立了基于啮齿类动物(大鼠)和非人灵长类动物(猕猴)的植入式脑机接口研究平台，以及面向临床康复领域的人机接口研究平台，在"复杂环境下的大鼠导航"、"基于 P300 的中文打字机"、"小动物双向的脑机接口系统"、"基于大鼠运动神经解码的神经控制"和"实时运动神经解码理论与方法"等项目的研究中取得了重要的突破。

2.1　大鼠导航

2006 年，我们研发了动物导航系统，利用植入实验动物感觉皮层和内侧前脑束的刺激电极和特定的刺激模式，动物可以获得虚拟的"触觉"和"奖赏"。通过自行研发的无线微刺激器的控制，实验动物可根据虚拟触觉的暗示和虚拟奖赏的指令完成左右转向和前行等动作[1]。在原有的工作基础上，2010 年我们结合操作性条件反射和直接诱发反应构建了混合控制方式，将停止功能引入到大鼠机器人之中，可在无约束的旷野环境中实现复杂的空间导航(如图 1 所示)[2]。

2.2　小动物双向脑机接口系统

我们利用嵌入式技术开发完成了基于 ARM9 处理器的无线低功耗的刺激-记录双向系统[3]。便携式的 PDA 不但可以控制大鼠背包上的刺激电路产生刺激脉冲，而且可以同时记录多通道的神经信号。该系统的高度便携性和集成性可用于一些复杂环境，完成自由活动动物在体脑电信号采集和特定神经核团激励(图 2)，为大脑等神经系统的工作机制研究提供了新的实用工具。

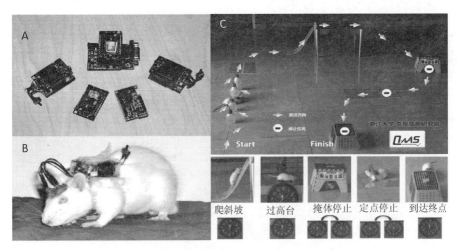

图1　复杂环境下的大鼠导航系统

A. 无线刺激背包；B. 大鼠机器人；C. 在无约束的旷野环境中，具有停止功能的大鼠
完成复杂导航任务

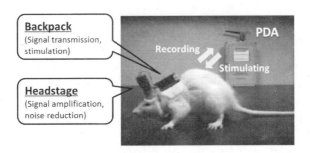

图2　刺激-记录双向小动物背包

2.3　运动神经解码的相关研究

"基于运动神经解码的脑机接口技术"是当前国际上脑机接口研究的前沿
热点。因此，为了深入研究运动皮层神经解码理论和方法，通过四年多的努
力，我们已经初步建立了基于啮齿类动物(大鼠)和非人灵长类动物(猕猴)两
个植入式脑机接口的动物研究平台，自主研发了多套可用于神经集群活动与
动物行为同步记录与分析的系统。

在大鼠研究平台上，我们利用自行研发的多通道微电极阵列可以长期稳
定地对大鼠运动皮层的神经信号进行记录(平均记录时间超过6个月)。在构
建了一系列效率高、抗噪能力强的神经锋电位检测分类算法的基础上，我们利
用传统的线性神经解码算法在离线(Off-line)的条件下实现了神经活动信号
对于动物前肢运动的预测，如图3所示[4]。

　　已有的研究表明,大脑的神经信号具有典型的非线性非稳态特性。为了适应大脑的非线性,我们构建了一种基于概率神经网络(PNN)的非线性神经解码算法,进一步提高了解码的准确度;同时,为了克服非线性算法计算量大的特点,我们利用并行计算的思想在 FPGA 对神经解码算法进行了优化,研究结果显示计算速度可提高将近 40 倍,提高了计算的实时性能,如图 4 所示[5];为了适应大脑的非平稳性,我们基于广义回归神经网络构建了一种时变解码模型,可以在神经解码的同时,利用新的神经数据对模型进行更新,从而实现了动态解码[6]。

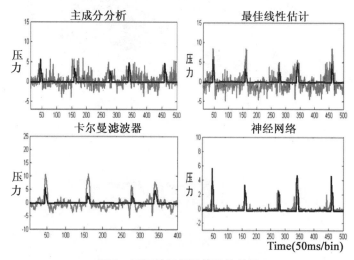

图 3　不同神经解码算法的效果
黑色为实际压杆上的压力值,灰色为神经信号预测的压力值

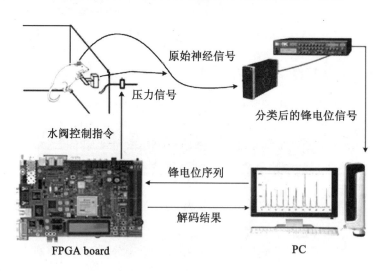

图 4　FPGA 实现的神经概率网络算法的实时神经解码系统

　　尽管目前绝大多数的神经信号解码主要依赖于神经元的锋电位信号，但是其他类型的神经信号是否也参与了信息的编码，它们携带的信息与锋电位信号所携带的信息又有什么区别呢？通过在大鼠运动皮层上同步记录到的锋电位信号和场电位信号，我们对两种信号的解码性能进行了研究[7]，结果显示，局部场电位信号同样携带了运动信息，两种信号的联合，可以进一步提高神经解码的准确度，增强脑机接口性能的稳定性，如图5所示。

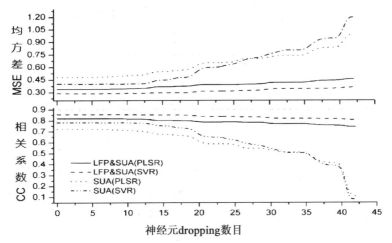

图5　场电位结合锋电位解码可以提高神经解码的准确度，增强脑机接口性能的稳定性。其中，**LFP—场电位信号，SUA—锋电位信号，PLSR—部分最小线性二乘拟合，SVR—支持向量回归**

　　经过在大鼠实验平台上近三年的研究和探索，我们从 2009 年下半年开始在猕猴研究平台上进一步开展植入式脑机接口研究，希望通过与人类更接近的猕猴，深入研究上肢的运动控制，特别是对于手指的精确运动控制。经过将近两年的摸索，目前已建立好相关的实验平台，已经分别在 3 只猴子的初级运动皮层和(或)背侧前运动皮层植入 96 通道的微电极阵列(图6)。利用经典的Center-Out 实验范式，我们已经在离线的条件下对手臂的大关节运动参数进行了解码，如图 7 所示[8]。下一步我们将重点对手指的运动参数进行解码，

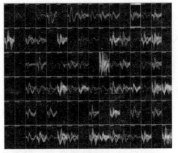

(a)植入微电极阵列的实验动物　　　　(b)猕猴初级运动皮层神经信号

图6　基于猕猴的植入式脑机接口平台

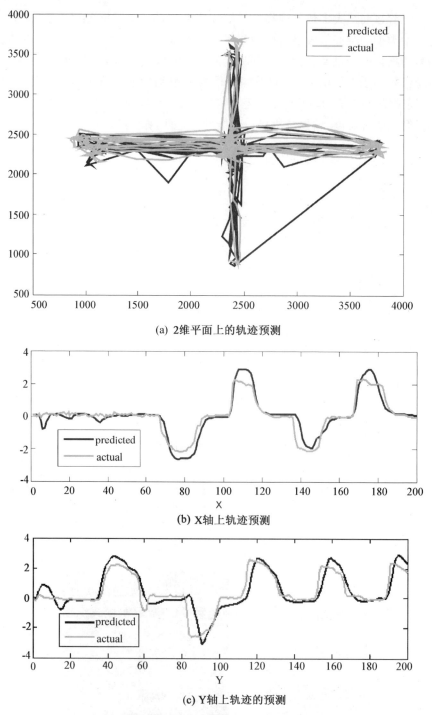

(a) 2维平面上的轨迹预测

(b) X轴上轨迹预测

(c) Y轴上轨迹的预测

图7　猴子上肢运动轨迹的神经解码。黑色实线 (predicted) 是神经解码的预测结果，
灰色实线 (actual) 是实际手腕运动的轨迹

为此我们还建立了手指的运动捕获装置,如图 8 所示。我们最终的目的是通过神经信号的解码实现对智能假手的控制,相关工作我们正在与意大利圣安娜高等研究大学的 Carrozza 教授领导的 ARTS Lab 开展密切的合作。

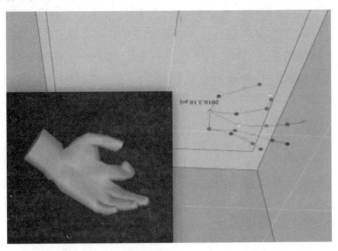

图 8　运动捕获的手指精细运动

3　总结

　　尽管过去的十年间植入式脑机接口的研究已经取得了引人注目的进展,但是在信息解析与交互等理论与关键技术方面仍面临巨大的挑战。我们在大鼠和猕猴上的相关工作和探索,进一步使我们认识到,一个实用的脑机接口系统依赖于以下两个关键性科学问题:① 动态高通量神经集群信息的实时高效解析;② 基于多模态反馈的自适应脑机交互。两者互为依托,准确、实时的信息解析有助于交互的完成,而交互的实现将有利于生物体神经信息的自然表征。第一个关键问题的解决不仅取决于多通道神经集群信息的约简等高通量神经信息预处理的关键技术,还取决于实时、动态的神经集群信息解析的关键技术,需要在神经集群信息处理、模式识别等技术上有所创新。第二个关键问题主要涉及多模态反馈技术和外部设备的智能控制策略研究,实现各种模态的反馈以及生物体与外部设备之间的实时、精确控制,需要在信息感知与融合、智能控制等关键技术上有所突破。此外,以上两个问题的解决还有赖于我们在神经科学方面对大脑的工作机制不断的深入研究,生物智能的研究不仅能在理论上为脑机接口研究提供指导,而且能够进一步帮助我们提高现有脑机接口的性能,甚至在不久的将来还可能实现生物智能和人工智能的融合,为我们的工作、学习和生活带来革命性的变革。

以脑机接口技术为代表的神经工程是当前多学科交叉研究的前沿,它的发展需要来自神经科学、信息科学和工程技术等多个领域研究人员在全球范围内进行广泛的交流和合作,形成知识互补和技术综合。最后,非常感谢我的团队和我的学生,所有的成绩是和你们努力的工作分不开的,也同样感谢我们的合作伙伴和那些给予我们帮助和指导的朋友们,能跟你们一起工作是我们莫大的荣幸。

参考文献

[1] Feng Z, *et al*. A remote control training system for rat navigation in complicated environment. *Journal of Zhejiang University-Science A*, 2007, 8(2): 323 - 330

[2] Lin J, *et al*. Using dlPAG-evoked immobile behavior in animal-robotics navigation: IEEE

[3] Ye X, *et al*. A portable telemetry system for brain stimulation and neuronal activity recording in freely behaving small animals. *J Neurosci Methods*, 2008, 174(2): 186 - 93

[4] Dai J, *et al*. Decoding of spatiotemporal firing pattern of neural ensemble in rat primary motor cortex. *SCIENCE CHINA Life Sciences*, 2009, 39(8): 736 - 745

[5] Yu Y, *et al*. Neural decoding based on probabilistic neural network. *J Zhejiang Univ Sci B*, 2010, 11(4): 298 - 306

[6] Zhou F, *et al*. Field-programmable gate array implementation of a probabilistic neural network for motor cortical decoding in rats. *J Neurosci Methods*, 2010, 185(2): 299 - 306

[7] Zhang S, *et al*. Decoding the nonstationary neural activity in motor cortex for brain machine interfaces. *International Journal of Imaging Systems and Technology*, 2011, 21(2): 158 - 164

[8] Zhang S, *et al*. A study on combining local field potential and single unit activity for better neural decoding. *International Journal of Imaging Systems and Technology*, 2011, 21(2): 165 - 172

[9] Zhang Q, *et al*. Building Brain-machine interfaces: from rat to monkey, in Asian Control Conference, ASCC, 2011(in press)

讲座人简介

Xiaoxiang Zheng received the Bache-lor degree in Radio Technology from Zhejiang U-niversity, Hangzhou, China, in 1968, M. S. and M. D. in Basic Medical Science from Tsukuba University, Japan, in 1984 and 1993.

Since 1993, she is Full Professor of Bio-medical Engineering, Zhejiang University where she has been a faculty member for more than 20 years. Since 1999 to 2005, she was Dean of College of Biomedical Engineer-ing & Instrument Science, Zhejiang University. Since October 2006, she is the Executive Director of Qiushi Academy for Advanced Studies, Zhejiang University. She is board member of the Chinese Medicine Research Commit-tee. She is member of Chinese Society of Biomedical Engineering. She had scientific and coordination responsibilities within more than 30 research pro-jects, funded under the Ministry of Science and Technology of China, the Ministry of Commerce of China and the National Natural Science Foundation of China. She is author of more than 300 scientific papers (more than 60 SCI papers and more than 60 EI papers). She also is a recipient of numerous pro-vincial and ministerial science or technology awards. Her primary research interests include microcirculation, physiology of cell, drug screening and neural engineering.

植入式神经接口和微型生物传感器的研究

Florian Solzbacher

(Electrical and Computer Engineering, University of
Utah and Blackrock Microsystems, Salt Lake City, UT 84112, USA)

摘要:微纳米制造与封装技术的不断进步,使我们有可能实现基于硅和聚合物的有线/无线集成系统,用于短期或长期的植入应用。本报告主要介绍植入式神经接口和微型生物传感器,这些系统可用于人或动物神经电生理和新陈代谢参数的采集,进而推动神经科学的研究以及临床应用。报告详细描述了这两种设备的制作和植入过程,以及长期在体和离体使用时的性能。

关键词:神经接口;无线;Utah 电极;微型生物传感器;新陈代谢

The Study of Implantable Neural Interface and Micro-biosensors

Florian Solzbacher

(Electrical and Computer Engineering, University of Utah and
Blackrock Microsystems, Salt Lake City, UT 84112, USA)

Abstract:Recent advances in micro/nanofabrication and packaging technologies have allowed the integration of active wired and wireless systems on Silicon and polymer platforms for acute and chronic implantable applications. The presentation will introduce new devices and system concepts including the implantable neural interface and micro-biosensors forgathering neurophysiological and metabolic parameters of human and animal. These systems have the potential to promote neuroscience research and clinical applications. Device fabrication, implantation, and long-term performance in vitro and in vivo are demonstrated in detail.

Keywords：Neural interface，Wireless，Urah electrode，Micro-biosensors，Metabolism

1 引言

用于临床治疗的人体植入装置或设备一直是生物医学工程领域的重大挑战。植入设备的关键特性是生物相容性，不对人体造成伤害，并且能长期在人体内稳定工作。实现此目标需要进行大量研究与实验，尤其是针对神经系统开展专门研究。其中一个重要问题是怎样长期稳定有效地采集脑电信号，这主要涉及以下两个研究点。

1.1 微电极阵列

目前主要有四种微电极，包括微丝电极、玻璃锥状电极、刚性电极和柔性电极。微丝电极通常由超细金属丝加工制得（$12\sim50\mu$m），玻璃锥状电极主要通过加热毛细管拉制而成，刚性电极则是在硅基材料上采用金属沉积、光刻、反应离子刻蚀等工艺加工而成，具有特定形状与尺寸。这三种电极多是针型的，典型代表有 Utah 式针型微电极和 Michigan 式针型微电极。第四种电极，即柔性电极，以聚合物为基底材料，如有聚酰亚胺（Polyimide，PI）和聚对二甲苯（Parylene）。这些聚合物材料不仅具有很好的生物相容性和微加工工艺兼容性，而且具有良好的机械性能和介电性质。同样地，采用金属沉积、光刻、反应离子刻蚀等工艺步骤，可加工成特定形状与尺寸的微电极，除了有针式外形以外，还有箍形和筛形。

1.2 电极阵列采集到信号的传输问题

主流的信号传输方法是使用电缆线进行有线传输，但该方法对载体的运动有一定的影响，而且不利于临床应用，因而人们提出了无线传输的方法。除了 Michigan 和我们的无线传输方案，其他课题组的无线系统大都采用的是独立器件，体积大，能耗高且不能长期植入。在 Utah 电极的基础上，利用先进的纳米制造与封装技术，我们设计了集成的无线传输系统，适于长期使用。除了脑电信号的检测，许多其他生理标记的采集对临床治疗也有很大的意义。采用类似的技术集成电子器件并进行封装，可制造面向其他应用的小型装置，例如可用于神经科学、神经疾病、新陈代谢监测等。

2 研究无线神经接口系统

为了消除有线神经接口系统中电缆的影响,我们采用新型的集成与封装技术,成功设计了无线神经接口系统,可植入大脑或其他神经系统位置。整个无线神经接口系统由 100 通道的 Utah 微电极阵列(Utah Electrode Array,UEA)、专用集成电路(Application Specific Integrated Circuit,ASIC)、能量线圈和贴片电容几部分组成。对采样后的脑电进行放大、滤波等初步处理,无线传输至处理器,系统能量也通过无线方式供应。经过测试,整个系统可以长期稳定地工作。

Utah 针型微电极是由 Normann 等人利用微电子技术加工出的微型硅针阵列。在硅针尖端沉积铂金,并对剩余部分的硅表面进行氮化硅处理形成绝缘层。在 0.4mm×0.4mm 面积内可加工 100 个棱锥形的硅针阵列,长度为 1.5mm,基部大小为 0.06mm×0.06mm。硅针的长度可变,范围为 0.5mm 至1.5mm。

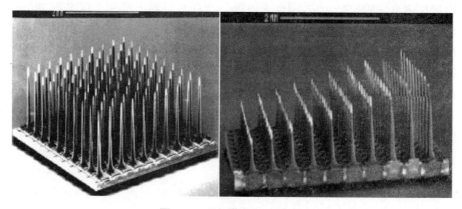

图 1 平面和梯面 Utah 电极

ASIC 包括放大器、信号处理单元、RF 发射端、能量接收单元和时钟恢复单元。采用 Au/Sn 回流焊接法,将 ASIC 倒装在 UEA 上。使用由 Ti/Pt/Au 构成的生物兼容材料优化凸点下金属化层(under bump metallization,UBM)。在 ASIC 和 UEA 之间填加材料,以改善系统的机械稳定性,并防止液体的进入。在聚酰亚胺基底上电镀金层实现无线能量耦合线圈,然后采用常规的金属陶瓷垫片和 SnCu 回流焊接法,将线圈集成到系统 ASIC 上。UEA 上贴片电容同样采用 SnCu 回流焊接法。

图 2　无线神经接口系统规格

3　微型生物传感器

我们研制的微型生物传感器,可用于监视新陈代谢的生理参数,如 pH、CO_2、葡萄糖与离子浓度等。传感器具有生物兼容性高与体积小的优点,可植入病人皮肤下,采集生理参数。现有固体膜片和穿孔膜片两类微型生物传感器,如图 3 所示,主要器件是水凝胶(smart hydrogels)和压阻式压力传感器。水凝胶是一种由交联的高分子聚合物网络组成的聚合物材料,根据外界环境的情况,吸收或释放水分子。微型生物传感器的基本原理是,环境溶液中的离子或目标分析物会穿过小孔,与水凝胶发生反应。当目标分析物的浓度发生变化时,导致外界液体的化学电势和渗透压发生相应的改变,水凝胶因吸收或释放水分子产生一定程度的膨胀或收缩。与水凝胶相连的压力传感器,实时记录其形变,从而分析环境溶液中离子或目标分析物浓度的变化。

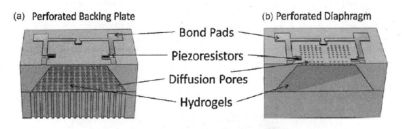

图 3　穿孔式和固体膜片式传感器

4　测试与应用

4.1　无线神经接口系统

将设备浸泡在典型的生理缓冲液(physiological buffer solutions)中进行性能测试实验。经观察发现,处理器接收端接收的 RF 信号保持稳定的频率,强度没有漂移,且信号噪声没有明显变化。当噪声升高到一定水平时,无额外动作电位产生,信号形状未出现任何形式的紊乱。结果表明整个系统比有线装置更有利于信号的传送。为了测试系统 ASIC 的长期工作稳定性,我们使用 Hermes C system 测试芯片噪声,并结合其他测试标准,发现 ASIC 能有效传输信号达 13 个月。

利用此无线神经接口,采集猴子大脑皮层的神经元发放信号(图 4)。将电极植入运动皮层,猴子可以自由行走。比较无线系统与有线系统采集的神经元发放信号后发现,无线系统能在猴子大脑稳定工作,为神经科学实验提供有效数据。

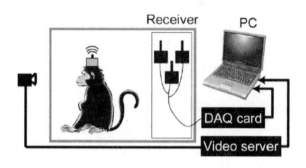

图 4　猴子在体实验

4.2　微型生物传感器

微型生物传感器的性能可进行离体测试,即将传感器放置在不同的生理缓冲溶液中,进行多次实验,得到不同目标分析物浓度曲线,典型例子有葡萄糖和果糖溶液(图 5)。实验中发现传感器对果糖的敏感度比葡萄糖要低,但是有类似的浓度变化曲线。传感器的在体测试是将其放置在牛血清中,采集不同目标分析物的浓度曲线。

图 5　PBS 缓冲液中测试

慢性测试结果显示我们研发的微型生物传感器能连续工作 400 小时,葡萄糖传感器能持续 5 天,在此期间,传感器灵敏度保持恒定。虽然采集到的数据存在一些漂移现象,但可通过电子补偿来克服。

实验结果表明微型生物传感器安全有效且能持续工作。

5 总结

安全有效的植入式设备是生物医学工程领域一项重要研究挑战,相关研究可以推动神经科学的研究和临床应用的发展。我们在这个领域进行了多年的研究,成功研制了植入式无线神经接口系统和微型生物传感器。无线神经接口系统消除电线对载体自由运动的限制,体积小且功耗低。微型传感器系统可以植入载体的皮肤下面,用于实时监测新陈代谢的生理参数,具有生物兼容性好且体积小等优点。大量的测试实验表明,这两种植入式设备能够长期稳定工作,具有优异的性能。

参考文献

[1] Chestek CA, *et al*. Hermes C: low-power wireless neural recording system for freely moving primates. *Neural Systems and Rehabilitation Engineering*, *IEEE Transactions on*, 2009, 17:330 - 338

[2] Kim S, *et al*. Integrated wireless neural interface based on the Utah electrode array. *Biomedical microdevices*, 2009, 11:453 - 466

[3] Orthner M, *et al*. Hydrogel based sensor arrays with perforated piezoresistive diaphragms for metabolic monitoring (in vitro). *Sensors and Actuators B: Chemical*, 2010, 145:807 - 816

[4] Sharma A, *et al*. Long term in vitro stability of fully integrated wireless neural interfaces based on Utah slant electrode array", *Applied physics letters*, 2010, 96:73702

[5] Tathireddy P, *et al*. Smart hydrogel based microsensing platform for continuous glucose monitoring. 2010:677 - 679

讲座人简介

Prof. Solzbacher is Director of the Utah Nanofabrication Laboratory, Co-Director of the Utah Nanotechnology Institute, President and Executive Chairman of Blackrock Microsystems and holds faculty appointments in E-

lectrical and Computer Engineering, Materials Science and Bioengineering at the University of Utah. His research focuses on harsh environment microsystems and materials, including implantable, wireless microsystems for biomedical and healthcare applications, and on high temperature and harsh environment compatible micro sensors. Prof. Solzbacher received his M. Sc. EE from the Technical University Berlin in 1997 and his Ph. D. from the Technical University Ilmenau in 2003. He is co-founder of several companies such as Blackrock Microsystems, Blackrock Sensors and First Sensor Technology. He was a board member and Chairman of the German Association for Sensor Technology AMA and of Sensor ＋ Test trade show and conference from 2001 until 2009, and serves on a number of company and public private partnership advisory boards and international conference steering committees. He is author of over 160 journal and conference publications, 5 book chapters and 16 pending patents.

人脑单神经元信号对外部设备的意念控制

Moran Cerf

(Computational Neuroscience, California Institute of Technology, the
Department of Neurosurgery, University of California, Los Angeles
and Stern School of Business, New York University)

摘要：认知对神经活动具有调节功能,比如大脑内侧颞叶的神经元能有选择性地响应某些特定的视觉对象。本文利用这种机制构建了一种新型脑机接口,在患者脑内植入电极,并要求其用意念对叠加的电脑合成图片进行控制,实现图像的淡入淡出功能。实验表明,患者可以迅速学会控制和调节不同半球和脑区内侧颞叶的神经元活动,通过增强或者减弱相关神经元的发放,实现对图片的控制与切换。这项工作的结果直接表明,人类可以有效调节自己的大脑神经元活动,从而实现对外部设备的控制。

关键词：单神经元解码;植入式;脑机接口

Projecting Thoughts to External Devices Using Single Neurons Recording in Human Brain

Moran Cerf

(Computational Neuroscience, California Institute of Technology , the
Department of Neurosurgery, University of California, Los Angeles
and Stern School of Business, New York University)

Abstract：Neurons in the medial temporal lobe (MTL) are selectively responsive to particular visual objects, and their activity is known to be modulated by cognitive effects. We have constructed a new brain-machine interface in which patients implanted with intracranial electrodes looking at superimposed computer images, when asked to make an image fade in or fade out, rapidly learnt to regulate neuronal activity of their MTL neurons in different subregions and hemispheres. They were able to increase the firing rate of certain cells while decreasing that of others and controlling the composite image content. This work provides direct evidence that humans can control the neuronal activity of their own visual neurons deep inside their own brain, and that such activity can be decoded to control devices.

Keywords：Single neurons recording, Implant, Brain-machine interface

1 引言

随着神经认知科学和脑机接口相关技术的不断发展,大脑和外部设备之间直接交互的手段也在不断丰富。通过神经解码的方式将大脑的主观意识和想法投射出来,从而用于外部设备的交互是当前国际上的研究热点。除了基于集群解码理论实现的控制方法外,另一种根据独立神经元解码的认知理论建立起来的新的脑机接口技术,同样可以实现大脑对于外部设备的直接控制。

对于癫痫病患者,需要通过颅内置入电极的方式,寻找和确定癫痫病灶的位置。通过对病灶位置的精确定位,医生将其切除,并尽量不影响脑部的正常功能和活动。电极植入后的几天里,医护人员一直采集病人的脑部数据,并等待或者诱发病人产生癫痫。当癫痫发作时,病灶附近位置的脑电活动会显示出明显的异常。在若干次癫痫发作后,我们通过观察分析神经数据,从而确定癫痫病灶的位置,然后通过手术将其切除,达到治疗的目的。借助这种方法,实验人员可以获得病人脑内的神经信号数据,并设计相关实验,研究大脑与外部设备的交互机制。需要注意的是,病人在常规病房中,不利于相关实验的进行。因为病人的脑部活动会受到各种干扰,比如外部噪音、医生护士们的随意进出房间等因素,所以我们首先要尽量去除实验环境中的各种干扰因素,从而获得比较"干净"的脑部数据。

颅内电极植入后,我们通过微电线(Mircowires)连接的微电极采集脑电数据。微电线的另一端链接脑电数据的放大和记录设备。电极深入大脑皮层内部,记录电极附近范围内的神经元动作电位的发放情况。由于基于单神经元解码理论,我们需要获取单个神经元的发放电情况,而微电极获得的数据可能包含了多个神经元同时发放时的叠加数据,所以我们首先要通过波形的判断对数据进行简单的分类,分离出单个神经元数据,用于后续的分析处理。

我们获取的神经信息并不是在大脑的表面,而是深入大脑内部获取的神经元发放情况,包括杏仁核、海马区、海马旁回等区域。这些区域主要是与人类的记忆、情感等有关。正是通过病人的记忆、意识等主观想法,我们得以将其转化为对外部设备的控制命令。

2 单神经元解码理论

大脑对于外部世界的感知具有不同的层次,以视觉系统为例,先从我们眼睛获得的视觉数据开始,通过不同脑区、不同层次神经元编码的协同工作,一直到更高层的认知行为,最终形成某种"概念"。比如说,当看到母亲的时候,首先是视觉上获得母亲的影像信息,经过更加高级的认知处理,最终形成一个关于"母亲"的概念。而根据单神经元解码理论,这个概念可以仅由一个神经元的数据就可以解码识别出来,我们的工作就是解码这些"概念"的神经数据,然后将它用于控制外部设备。

我们在实验中事先跟病人进行交流,了解病人的一些个人信息,比如最近去哪里出游,喜欢什么类型的电影、音乐,比较喜欢哪些明星等,然后根据这些数据挑选与这些内容相关的一百张图片。图片分成六个不同的组别,每一组对应病人个人信息中的一个独立的概念。然后实验人员将其反复地播放给病人看,时间大约持续 30 分钟左右。这里举一个例子,我们针对某个病人,给他看与他有关的一组图片,包括他个人喜欢的女演员哈利贝瑞、安妮斯顿、埃菲尔铁塔、克林顿、蜘蛛等。在病人反复观看这些图片的半个小时中,我们同时观察记录病人的脑电发放情况。根据分析杏仁核、海马等脑区的神经元发放情况,我们发现不同的神经元在看到不同组别的图片时会产生明显的差异。以海马区的某个神经元为例,我们发现该神经元在病人看到哈利贝瑞的图片时,会出现显著的一个锋电位发放的高峰。我们认为这个神经元与哈利贝瑞这个"概念"有关。

然后我们继续分析该神经元的发放情况,发现每次病人看到哈利贝瑞一组图片时,都会产生锋电位集中发放的情况;而对其他组的图片信息,该神经元没有明显的发放。我们把这个神经元叫做"哈利贝瑞"神经元。同时我们发现该神经元对应的不只是哈利贝瑞的某个图片,而是哈利贝瑞这个"概念"。每次我们给病人看哈利贝瑞的图片,包括她的头像、她的素描画,该神经元都会产生集中的锋电位发放。我们也发现当给病人看的图片包括另一个女人的猫女形象时(哈利贝瑞曾经出演过猫女的电影),病人会意识到这个人和哈利贝瑞有关,所以神经元也会集中发放,甚至病人看到"哈利贝瑞"这段文字的时候,该神经元也会发放,从而我们认为每次病人意识到和哈利贝瑞这个概念相关的时候,该神经元都会集中发放锋电位,即该神经元编码了"哈利贝瑞"这个概念,从而我们证实了单神经元解码理论,即通过对单个神经元的活动分析,可以解码出某个独立的概念。

这个理论在其他多名病人身上也得到了验证。我们给病人看不同的视频片段,同时分析其脑部某个神经元的锋电位发放情况。视频内容包括:好莱坞的介绍片,马丁路德·金的演讲视频,《泰坦尼克》电影片段,麦当娜的 MV,动画片《辛普森一家》,玛丽莲·梦露的电影,麦克·乔丹等不同的主题。以其中一个病人为例,我们找到该病人的某个神经元,在看到《辛普森一家》的相关影像资料时会集中发放。之后我们要求病人回忆刚刚看到的片段,并告诉我们他回忆的片段是什么,同时持续记录该神经元发放电情况。当病人回忆到《辛普森一家》时,神经元会提前发放,之后几秒钟病人才告诉我们他回忆的内容。研究表明,病人的某单个神经元不但会针对某个概念发放,而且在病人闭上眼睛想象这个概念的时候,神经元也会出现锋电位集中发放,而且病人在主观上意识到回忆起某个概念之前,与该概念相关的神经元已经出现了明显的发放高峰。

我们进一步实验,要求病人看一组不同的图片,当病人看到一张人脸图片时,同时会听到一声频率较高的提示音;而病人看到一个物品(如棒球)时,则听到频率低的声音。经过一段时间的训练,病人能够分别将人脸图片和物体图片对应到高频和低频的声音。之后我们要病人闭上眼睛,仅让其听到不同频率的声音,我们发现神经元在高低频率不同的声音提示下,也对应不同的概念(人脸或者物品),明显有不同的发放模式。这进一步证明了单神经元解码出的内容,不仅仅是视觉信息,而是基于概念的信息。

3 单神经元控制外部设备

根据上面的实验和理论,我们将大脑中单个神经元针对某个特定概念的活动情况解码出来,用以实现对外部设备的控制。在新的实验中,我们将病人安置在安静的实验环境中,通过脑电记录系统连接到病人的脑部电极,实时观测病人的锋电位发放情况。

实验开始后,我们给病人看四组图片,包括玛丽莲·梦露(演员)、大威廉姆斯(网球运动员)、迈克尔·杰克逊(歌手)、约什布罗林(演员)。图片的选择和建立与上面提到的实验相似,是通过跟病人事先交流选取其比较感兴趣的内容。我们记录下病人看到不同图片时候的脑电情况,然后要病人通过主动想象这些图片的内容,从而主动控制与该概念相关的神经元发放电频率和强度,通过解码将其反应出来,从而控制外部系统。举一个例子,被试的一位女病人,我们通过大量实验和数据分析得到四组不同图片时四个不同的动作电位发放显著的神经元，即找到针对每个概念编码的神经元。在左侧海马旁回区我们找到一个神经元在每次看到玛丽莲·梦露图片的时候会集中发放；而在右侧海马区，某个神经元在看到约什布罗林图片的时候会出现发放情况。

之后，我们将两张图片整合在一起，图片呈现出两个不同人物叠加在一起的状态。然后实验人员要求病人根据我们提示的目标图片，主动想象和该目标相关的内容，从而提高与其对应的神经元的活动频率，而两个神经元的活动频率通过数据解码出来，反应为在叠加图片中对应内容的显示比例。实验结果是，当病人主动想象某个图片内容时，对应的图片就逐渐清晰，而另一张则逐渐模糊，最终显示出完整的目标图片。

需要注意的是，这个神经元并不是唯一的反应某个概念的神经元，实际上有一组神经元的发放都与某个概念有关。这个实验需要一定的时间，使病人学会如何用脑电控制图片。开始时正确率只有 30% 左右，大约几分钟后，有四到五个神经元可以完美地进行解码。我们比较了不同脑区的神经元解码，在海马区、杏仁核等脑区的神经元表现更好。以上述女病人为例，经过大约 20min 的训练学习过程，病人对于玛丽莲·梦露图片的显示控制可以达到百分之百的正确率，而对约什布罗林图片的控制也可以达到 90% 以上，证明通过该方法可以实现大脑对于外部系统的直接有效控制。

深入的研究发现某些病人可以非常好地控制自己神经元的发放程度，甚至可以精确控制其发放频率。为此我们也设计了一组实验。实验中针对的是病人的"里根"神经元，该神经元与美国前总统里根的概念相关。我们设计一个游戏，病人的神经元活动反应为一个飞机，当病人想象里根相关的内容时，神经元发放强烈，飞机上升；当病人不再想象其概念，神经元不发放，飞机下降。我们设计飞机随着时间会遇到不同高度的障碍物，而提示病人通过控制神经元发放，控制飞机绕过障碍物。我们发现在一段时间的学习适应过程后，病人可以很好地控制飞机绕过障碍。但也发现有时在没有障碍的情况下，飞机也会主动上升。

我们根据多组实验统计发现，当病人自主地控制神经元发放时（即病人自发控制某个神经元发放），比被动要求控制的时候表现更好（即病人在某个时刻被要求控制神经元的发放）。最近我们又设计了一组新的实验：患者在床上面对电脑，屏幕上显示一钟面，钟面的指针会不停地转。被试者将食指放在键盘上等待按键。指针跑两圈后，被试者可以自由决定什么时候按键，一旦他们按键，指针就停下来。然后他们将指针调回到他们有按键意念时指针所指的位置。此时我们得到两个不同的时刻，一个是病人做出主观决定，试图将指针停下的时间 W，一个是指针停下的时间 P。

我们发现前额叶皮层内单个神经元在病人做出要停止钟表的决定的时刻（时间 W）之前大约 1500 毫秒，会产生明显的集中发放。我们找到了多个神经元,在病人做出决策之前,会有明显的发放过程。这就是说,从病人做出主观决策并且自己意识到之前,我们已经可以在神经元的层面上看到该决策行为,即病人的主观决定可以被提前预测。我们下一步会将其引入脑机接口的相关研究中来,从而提高脑机接口的效率。

4 结论

通过单神经元解码理论的实用化,我们建立了一种新的神经元解码控制思路和方法,可以实现对外部设备的有效控制。同时我们也发现了可以有效预测病人在决策时的主观意识。相比其他脑机接口中的"解码-控制"方法,该方法相对较简单,学习过程较短。但是由于该方法需要被试者将针对某个概念的意识转换为控制意图,实际上实现了一种间接的控制过程,所以不适用于其他动物。同时单神经元解码的维度较低,无法实现精确复杂的控制行为(如自由度较高的机械臂),影响了其在运动控制上的应用。

参考文献

[1] Quian QR *et al*. Invariant visual representation by single neurons in the human brain. *Nature*, 2005, 435: 1102 – 1107

[2] Cerf M, Thiruvengadam N, Mormann F, *et al*. Online volutary control of human temporal lobe neurons. *Nature*, 2010, 467:1104 – 1108

讲座人简介

Dr. Moran Cerf is neuroscientist at the California Institute of Technology (" Caltech "), UCLA department of neurosurgery and New York University. Dr. Cerf studies the neural basis of consciousness, and the ability to decode subjects' thoughts in real time. His research focuses on understanding the neural mechanisms of consciousness and free using direct recording of single neurons from the brains of patients undergoing brain surgery. Dr. Cerf completed his Ph. D at Caltech in computational neuroscience, and holds an MA in Philosophy of Science and a B. Sc in Physics, both from the Tel-Aviv University. Prior to his career as a scientist, Dr. Cerf worked as a hacker — breaking into banks and financial institutes, an air pilot and an inventor. Dr. Cerf currently holds a faculty position at the American Film Institute, teaching screen-writing, and is currently the winner of the U. S Moth story-telling competition.

Cognitive Computation: The Ersatz Brain Project

James A. Anderson[1], Paul Allopenna[1], Gerald S. Guralnik[2],
Daniel Ferrante[2], and John A. Santini, Jr[3]

1. Department of Cognitive, Linguistic and Psychological Sciences, Brown University, Providence, Rhode Island, 02912, USA, James_Anderson @ brown. edu
2. Department of Physics, Brown University, Providence, Rhode Island, 02912, USA
3. Alion Science and Technology, 240 Oral School Road, Mystic CT, USA

The Ersatz Brain Project is an attempt to develop programming techniques and software applications for a brain-like computing system. The brain-like hardware architecture design is based on a few ideas taken from the anatomy of mammalian neo-cortex. In common with other such attempts it is based on a massively parallel, two-dimensional array of CPUs and their associated memory. The design used in this project (1) uses an approximation to cortical computation called the **network of networks** which holds that the basic computing unit in the cortex is not a single neuron but groups of neurons working together in attractor networks; (2) assumes connections and representations in cortex are sparse; (3) scales in a natural way from small groups of neurons to the operation and integration of entire cortical regions. The resulting hardware computes using techniques such as local data movement, temporal coincidence, formation of discrete "module assemblies", and the topographic arrangement of the data on the array. Software for such a system becomes a curious blend of techniques, some reminiscent of analog computers and some of a more familiar, more discrete kind. Several illustrative software applications especially suitable for this architecture will be discussed. Simple example operations can recognize identity, symmetry, items from a short learned list and can implement some kinds of useful spatial transformations of the representation for a stimulus domain.

1 Introduction

Carver Mead, emeritus Professor of Engineering at Caltech, winner of the National Medal of Technology, a legend in semiconductor design, has commented: "Listen to the technology; find out what it's telling you. " The technology is starting to talk loudly, and it is telling us two things:

First, computer software applications are now reaching into new areas, areas of thought and intellect that humans consider uniquely our own.

Second, computer hardware is reaching fundamental physical limits on how fast the individual electronic devices that comprise computers can go.

At first these seem like unrelated developments. What do computer hardware limits have to do with making intelligent software? The Ersatz Project suggests that the two offer an opportunity to be joined in an interesting, perhaps unexpected way.

We claim that looming problems with computer hardware can be helped by looking at the way the human brain does things. And, perhaps, brain-like computers will be the design best suited for human-like applications. Most wonderful, most thought provoking, and perhaps most unsettling would be making a silicon device that not only acted intelligent, it might even be intelligent because it works on the same principles that our brain does. Therefore these two events, seemingly so different, when working together may create a new synthesis affecting computer science: building a truly brain like computer (in new hardware) to compute human-like cognitive applications (in new software).

1. 1 Cognitive Software

For my thoughts are not your thoughts, neither are your ways my ways, saith the LORD. **Isaiah 55:8**

Brains and computers are different in basic hardware and how they operate on information.

Computer hardware is blazingly fast. Even an inexpensive home computer contains a Central Processing Unit [CPU] , the device that actually runs the programs, that can execute a billion separate operations, one after the other, in one second. Computer hardware is reliable and never makes errors. Computer hardware works in a world of ones and zeros. But computers work on information in small quantities, computer words: depending on the computer and its age, 8, 16, 32, or 64 bits at a time.

Brain "hardware" is glacially slow in comparison. The basic nerve cells — neurons — rarely operate faster than 1,000 times a second, a million times slower than a silicon CPU. There are a whole series of essential biological mechanisms that make nerve cells noisy. Nerve cells are affected by many malign influences, from bad biochemicals, to mechanical shock, to viruses and bacteria. But brain hardware works in a continuous world, that is, instead of only one's and zeros, neurons can signal all the values between zero and their fastest response rate. The cerebral cortex processes information in huge chunks. Instead of 64 bits at a time, ten billion nerve cells can be working on the same problem at the same time.

The hardware is so different that it is surprising that anyone ever thought they worked in the same way, but a lot of smart people did. The term "Artificial Intelligence" [AI] was first used at a famous summer long gathering at Dartmouth in 1956. Most of those who thought about the problem of smart machines were there for at least part of the summer. Their goal was to mimic human intelligence with a machine:

"AI's founders were profoundly optimistic about the future of the new field: Herbert Simon predicted that "machines will be capable, within twenty years, of doing any work a man can do" and Marvin Minsky agreed, writing that "within a generation ... the problem of creating 'artificial intelligence' will substantially be solved" [1].

Alas, such was not to be.

There was a consensus at that time that intelligent systems were forced by some unspecified law of nature to follow common universal rules of reasoning, and thinking. As a convenient consequence, if you understood machine intelligence well enough, you didn't have to spend time learning the details of human intelligence because they were the same. Since the Dartmouth participants were mathematicians, philosophers, computer scientists and engineers, they assumed intelligence in its general form worked how they thought, or, more accurately, how they thought they thought. Computers were good at logic and well structured analysis applied to well-structured problems, as were they. Therefore, the proposed match between human and machine intelligence was natural.

Unfortunately, as rapidly became apparent, humans did not use logic, reasoning, or formal analysis very often, and, when they did, they were bad at it. The first wave of AI research was a hugely expensive failure (of course, with lavish government support), though an instructive one. The great generalization of intelligence to all intelligent entities did not appear. The successes of AI, and there were some, were to very specific problems: specialized medical issues like diagnosis and drug interactions, configuring computers, doing some kinds of data analysis, and providing artificial players for video games and internet card games to compete with humans.

But, when humans and machines approach the same problem, unlike the hopes of early AI, they usually do so completely differently.

By chance, The *New York Times* for February 17, 2011 contains two important examples of just how different humans and machines are when "they" work on the same problem. On the front page is an article by John Markoff on a program called "Watson", developed by IBM that beat two human champions at the TV game "Jeopardy!" over a three-day contest. On page 21 of the same day's paper is a brief review of a previous IBM project, "Deep Blue", which beat the human world chess champion, Gary Kasparov in a six game match in 1997.

What is so striking about these two IBM projects is that the computer excelled in domains that humans had previously considered their own, as the founders of Artificial Intelligence hoped would be the case. But it also shows something else. The machines performed these tasks in a very different way than a human would.

Let us look at computer chess. Chess was considered by the original AI researchers to be a paradigmatic example of a human skill that could be performed by a machine. There has been extensive work on computer chess since the 1950's. The game is a good test problem because the rules of chess are clear, chance plays no role, the board positions are presented to both sides and there is a huge body of human experience with chess. If we had access to a computer of truly cosmic size, the outcome of the game would be totally determined before a move was made. It would know that if the best possible moves were made by both players, whether white would win, black would win, or whether the game would be a draw. Our computers are still far less powerful than that. But even so, modern computers solve chess largely through brute force.

When IBM's "Deep Blue" beat the world chess champion, Gary Kasparov it did so by using raw computer power. Aside from the initial few moves, based on exhaustively analyzed historical examples — the "book", based on 700,000 grandmaster games — Deep Blue, in common with almost all chess programs simply starts with the current board position and looks at the possible legal moves it can make and the most effective responses the opponent can make as far into the future as time allows. (Chess tournaments are run with a strict time limit.) The computer then chooses the best possible move from all the ones they examined.

As computers get faster, they can look deeper and deeper into future moves. Millions more moves can be evaluated. Deep Bluein 1997 could evaluate over 200,000,000 board positions per second and look six or eight moves into the future, and, in some specialized situations, the end game, for example, up to 20 moves ahead.

So the increase in success of computer chess programs really reflected the general increase in hardware capability of the industry. Deep Blue also had some hardware specializations added specifically for chess. Deep Blue used thirty high performance CPU's enhanced with 480 "chess chips" designed to efficiently code and manipulate board positions. [2]

1. 2   Brain Meets Machine

Physically, Deep Blue is an imposing six foot high chunk of hardware: two relay racks full of electronics, hundreds of pounds of metal. But consider the human chess player. A human brain weighs about 3 pounds (1.5 kg) and is about 1200 cubic centimeters in size. The energy consumption of the brain is about 25 watts. Yet this small, energy efficient structure is roughly at parity of processing power with a highly specialized, heavy, energy consuming bank of electronics.

Most striking, instead of evaluating hundreds of millions of future board positions, the human brain can only evaluate a few. Estimates of human masters based on experiment and interviews suggest they evaluate around fifty future board positions. Yet, obviously, they must somehow e-valuate the **right** board positions to get rough equality in power with a device evaluating millions of positions. Somehow our brains can use an effective combination of memory, perception, and intuition to achieve a result of comparable power to a computer but in a very different way.

After the game, Deep Blue was decommissioned and played no more matches. One of its two banks of electronics is in the Smithsonian Institu-tion in Washington, DC.

After the game, Kasparov walked off the stage without walking into the walls. Deep Blue stayed put. Kasparov walked, talked to reporters, ordered dinner, and has played in many further matches, although with humans. He retired from playing chess in 2005 to write and do political work including a potential run for President of Russia, a career path unlikely to be followed by Deep Blue. Deep Blue (the computer) might have won the tournament but Kasparov (the human) is far more versatile and can do a lot of things other than chess, all with a mobile, low power consumption CPU.

We note in conclusion that a chess grand master evaluates two million times fewer future board positions than Deep Blue but is roughly equivalent in power. Making a fast machine that could use such striking, efficient, powerful brain-like algorithms would be a formidable machine indeed.

1.3 Jeopardy

In 2011 the IBM program Watson defeated two past Jeopardy! quiz shows champions on network TV. Jeopardy! questions are largely a test of memory combined with effective used by human players of heuristics (good guesses), inference and the ability to understand bad puns.

Watson "practiced" for the tournament by learning immense amounts of fac-tual information and analyzing an enormous number of examples of text. Clever inference techniques then combined and searched this information on the fly to an-swer the questions. Watson had effective natural language understanding ability and was able to estimate its level of confidence in its answers.

Watson used 10 six foot racks of computer equipment containing 2,880 CPU's and 16 terabytes of memory, at least a ton of expensive electronics[3]. One estimate is that Watson can process 1,000,000 book equivalents per sec-ond. Again, the human players, with the three pound brain, do not have the luxury of extended search but must come up with answers quickly based on effective use of memory, association, constraints, inference and intuition.

This observation suggests humanscan creatively and quickly use memory to draw new conclusions to questions they have never seen before. Human memory is not in the form of vast quantities of isolated facts, but in the form of extremely effective, rapidly reconfigured generalizations.

Because of human speed and capacity limitations, most detail is not stored at all in memory, but only the most valuable generalizations and abstractions. In engineering, this kind of storing of just the right information for a task and discarding the rest is called sometimes "destructive data compression". Computers work quite differently. They can keep in storage many more details and extract the gist if needed for a specific task.

Here we suggest that the way humans work with memory and its interactions with information coming in from the senses is very different than a computer. Humans, with slow and inaccurate hardware can perform with comparable speed and power to fast, accurate silicon for many important tasks. Therefore joining human software techniques to a machine could produce a system of great power and efficiency. In addition, it might be an artificial system that might work well with humans since it would use the same basic approaches to dealing with knowledge. We might build an artificial "super friend", a silicon guide and confidant that would stand between human and computer ways of thinking and interpret one to the other.

1.4 Cognitive Hardware

Consider the hardware used by brains and machines. Brains have been around for hundreds of millions of years and are not going to change quickly. But since 1945 and the construction of the first computers, the American ENIAC and the British Colossus, digital computers have gotten faster and more capable in every way. But as their computing capabilities got bigger and bigger, physically their electronic components got smaller and smaller. The semiconductor chip industry is now manufacturing chips with individual component sizes well under that of a wavelength of light or the components found in biological cells. The "gates", a critical part of the transistor that allows them to turn on or off reliably are a few tens of atoms thick.

But physical hardware is reaching a technological limitation. In the recent past, small predictably got smaller with time. More components on a chip mean more computer power. In the computer industry, this trend follows what is called "Moore's Law," named after Gordon Moore, a co-founder of Intel. Computer speed and power has been doubling roughly every two years because more and more, smaller and smaller, devices can be packed onto a computer chip. Remarkably, Moore's Law — not really a Law but an observation — has been faithfully followed for almost fifty years, since Moore proposed it in 1965.

Besides getting more of devices on a chip, smaller devices are also faster devices. However, it is now widely believed that the increased computer power described and predicted by Moore's law will be coming to an end soon. Devices cannot become much smaller and the reasons for this are due to unavoidable fundamental laws of physics: smaller devices beyond a certain point — which we are now approaching — become unreliable due to quantum mechanics, the science of the extremely small.

One quantum mechanical effect is "tunneling". Even a very good insulator if it is thin enough will have some electrons pass through it. When transistors are as small as they are becoming, the probability of electrons tunneling through an insulator becomes significant, and gives rise to excess heat. More serious, tunneling and other quantum effects, predict that very small computer elements, of a size we are now approaching, could simply spontaneously change state, say from TRUE to FALSE, ON to OFF, or 0 to 1. This event would be a disaster for computer reliability. It would be hard to sell a computer that gave different, random answers to the same problem. Whether Moore's law fails next year, or in two decades, does not matter. The end is in sight.

1. 5   Hardware and Software Synthesis

Out of this nettle, danger, we pluck this flower, safety. **Shakespeare, Henry IV. , Act ii. , Sc. 3.**

What to do? One short term response of the computer industry has been to observe that if one computer is fast, two computers computing at the same time might be twice as fast, that is, go to parallelism. Chip makers are now shipping in large quantities chips built with multiple "cores", that is, two, four, eight or more CPUs on a single chip. Multi-core CPU's have now become nearly universal even for inexpensive home computers.

Parallelism has been studied by computer scientists for a generation because of its clear potential advantages in speed and hardware. Unfortunately, parallelism really works well not only for a restricted set of problems, most notably the graphics seen in movies and video games but also for some important but specialized scientific and engineering applications. Most real programs have both a serial part where steps must be done one after another (waiting for a result before proceeding), and a parallel part (say, moving images in a block around in a video game). This situation is described by "Amdahl's Law", named for Gene Amdahl, an influential computer architect and founder of Amdahl Corporation, now part of Fujitsu. Amdahl pointed out that even if the parallel parts of the program were infinitely fast, the program would still take finite time to run because of the need to wait for the serial parts of the program that must be done in sequence to finish.

But this discussion suggests that a golden opportunity appears if cognitive operations can be performed on parallel computers. The end of Moore's law is telling us, "If you can't get smaller, to get more powerful you must get help from your silicon brothers. " But we have had trouble making this team pull together. Maybe instead of making the new team play the same old game, which has not worked so well, we can use the new team to play a new game that they can win because they are well designed to play it.

One future trend   hardware limitations — can be addressed through parallel computers. And if these parallel computers can be applied successfully to the new human cognitive applications, everyone will win.

Human brains evolved a particular structure to perform certain classes of operations that are important to us as humans. The human cerebral cortex is very slow, but very highly parallel. A traditional computer is very fast, but does one simple operation after another in serial.

At present, and perhaps surprisingly, brains and computers seem to be of comparable power for a number of important "cognitive" operations such as chess or Jeopardy! Humans are far more versatile.

One future way intelligent machine technology might evolve is to join the two approaches. Both have strengths. Humans are strong at putting large amounts of information together, recognizing and working with patterns, forming simplifications from complexity, and are exceedingly flexible. Computers are precise, but inflexible, and work tirelessly with the finest detail in a way that humans find impossible. Merging the virtues of human computation and computer computation would provide a synthesis that would let machines work on difficult human problems.

Next, we discuss one approach to doing exactly this: the Ersatz Brain.

2   Essentials of the Ersatz Approach

The human brain is composed of on the order of 10^{10} neurons, connected together with at least 10^{14} connections between neurons. These numbers are likely to be underestimates. Biological neurons and their connections are extremely complex electrochemical structures that require substantial computer power to model even in poor approximations. The more realistic the neuron approximation, the smaller is the network that can be modeled. Worse, there is very strong evidence that **a bigger brain is a better brain**, thereby increasing greatly computational demands if biology is followed closely. We need good approximations to build a practical brain-like computer.

2. 1   The Ersatz Cortical Computing Module and the Network of Networks

Received wisdom has it that neurons are the basic computational units of the brain. However the Ersatz Brain Project is based on a different assumption. We will use the **Network of Networks** [**NofN**] approximation to structure the hardware and to reduce the number of connections required.

Generic Network of Networks Module

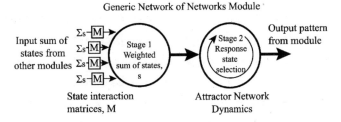

Figure   1

We assume that the basic neural computing units are not neurons, but small (perhaps $10^3 - 10^4$ neurons) attractor networks, that is, non-linear networks (**modules**) whose behavior is dominated by their attractor states that may be built in or acquired through learning. Basing computation on module attractor states — that is, on intermediate level structure — and not directly on the activities of single neurons reduces the dimensionality of the

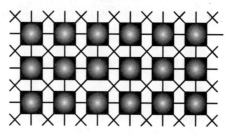

Network of Networks Modular Architecture

Figure　2

system, allows a degree of intrinsic noise immunity, and allows interactions between networks to be approximated as interactions between attractor states. Interactions between modules are similar to the generic neural net unit except scalar connection strengths are replaced by state interaction matrices. The state interaction matrix gives the effect of an attractor state in one module upon attractor states in a module connected to it. Because attractors are derived from neuron responses, it is potentially possible to merge neuron-based preprocessing with attractor dynamics. The basic Network of Networks system is composed of very many of these basic modules arranged in a two-dimensional array. Some of the basic ideas behind the Ersatz brain were developed in conjunction with Sutton[4,5].

2.2　Cortical Columns

The most likely physiological candidate for the basic component of a modular network is the cortical column. Cerebral cortex is a large two-dimensional layered sheet, with a repetitive structure. One of its most prominent anatomical features is the presence of what are called columns, local groups of cells oriented perpendicular to the cortical surface. There are several types of columns present at different spatial scales. Mountcastle summarized the situation as:"The basic unit of cortical operation is the minicolumn ... It contains of the order of 80 - 100 neurons, except in the primate striate cortex, where the number is more than doubled. The minicolumn measures of the order of 40 - 50 μm in transverse diameter, separated from adjacent minicolumns by vertical cell-sparse zones which vary in size in different cortical areas. Each minicolumn has all cortical phenotypes, and each has several output channels. ... By the 26th gestational week the human neocortex is composed of a large number of minicolumns in parallel vertical arrays. " [6]

Minicolumns form a biologically determined structure of stable size, form and universal occurrence. What are often called "columns" in the literature are collections of minicolumns that seem to form functional units. Probably the best-known examples of functional columns are the orientation columns in V1, primary visual cortex. Clusters of minicolumns make up functional columns:

Mountcastle continues, "Cortical columns are formed by the binding together of many minicolumns by common input and short range horizontal connections. The number of minicolumns per column varies probably because of variation in size of the cell sparse inter-minicolumnar zones; the number varies between 50 and 80. Long-range, intracortical projections link columns with similar functional properties. Columns vary between 300 and 500 μm in transverse diameter, and do not differ significantly in size between brains that may vary in size over three orders of magnitude ... " [6]

If we assume there are 100 neurons per minicolumn, and roughly 80 minicolumns per functional column, this suggests there are roughly 8,000 neurons in a column.

2.3 Connectivity

Besides modular structure, an important observation about the brain in general that strongly influences how it works is its very sparse connectivity between neurons. Although a given neuron in cortex may have on the order of 100,000 synapses, there are more than 10^{10} neurons in the brain. Therefore, the fractional connectivity is very low; for the previous numbers it is 0.001 per cent, even if every synapse connects to a different cell. Connections are expensive biologically since they take up space, use energy, and are hard to wire up correctly. The connections are precious and their pattern of connection must be under tight control. This observation puts severe constraints on the structure of large-scale brain models. One implication of expensive connections is that short local connections which are relatively cheap compared to longer range ones. The cortical approximation we will discuss makes extensive use of local connections for computation in addition to the sparse, accurately targeted long-range connections.

2. 4   Interactions between Modules

Let us discuss in a little more detail how to analyze interactions between small groups of modules. The attractor model we will use is the BSB network[7] because it is simple to analyze using the eigenvectors and eigenvalues of its local connections.

The BSB model was proposed several years ago as a simple feedback nonlinear neural network. Its dynamics broke conveniently into a linear and a non-linear part. The analysis assumed it was a recurrent feedback network. An input pattern, $\mathbf{f}$, appears on an interconnected group of neurons, say from a sensory input. There is vector feedback through a connection matrix, $\mathbf{A}$, weighted by a constant α and an inhibitory decay constant, γ, with amount of inhibition a function of the amplitude of the activity. The state of the system is $\mathbf{x}(t)$.

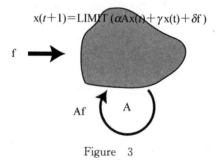

$$x(t+1)=\text{LIMIT}(\alpha Ax(t)+\gamma x(t)+\delta f)$$

f

Af   A

Figure   3

The system is linear up to the point where the LIMIT operation starts to operate. $\text{LIMIT}(\mathbf{x}(t))$ is a hard limiter with an upper and lower threshold. Sometimes it is useful to maintain the outside input $\mathbf{f}(0)$ at some level; sometimes it is useful to remove the outside input. The constant δ performs this function.

The basic algorithm for BSB in a single module is:

$$\mathbf{x}(t+1)=\text{LIMIT}(\alpha\mathbf{A}\mathbf{x}(t)+\gamma\mathbf{x}(t)+\delta\mathbf{f}(0)). \tag{1}$$

A more realistic version of module dynamics will incorporate lateral interactions between modules, as shown in the NofN illustration and overall general gain control terms. Note that activity is a continuous scalar value. In a real system, action potentials would be used. Use of action potentials means that low levels of activity become effectively zero since spikes would occur so infrequently. In the nonlinear BSB network with growing activity, the state of the system will reach an attractor state based on the LIMIT function, usually the corner of a hypercube of limits. In practice, if **f** is an eigenvector the final BSB attractor state is close to the direction of **f**.

Activity can increase without bound or go to zero. The transition from increasing activity to decreasing activity is under control of α, γ and, the eigenvalues of **A**. These parameters provide a way of controlling network behavior.

Let us consider the implications of connections from other structures. In particular, we know two relevant facts about cortex: (1) one cortical region projects to others and, (2) there are back projections from the target regions to the first region that are comparable in size to the upward projections.

$$x(1)=\text{LIMIT }(\alpha\lambda+\varepsilon+\gamma)f \qquad y(1)=\text{LIMIT }(\beta\mu+\eta+\gamma)g$$

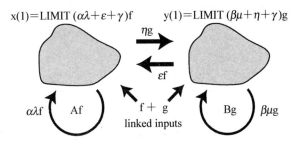

ηg

εf

$\alpha\lambda f \quad Af$

$f + g$
linked inputs

$Bg \quad \beta\mu g$

Figure 4

We can also propose a computational mechanism for binding together a multimodule input pattern using local connections. If two modules are driven by two simultaneously presented patterns, **f** and **g**, associative links between **f** and **g** can be formed, increasing the gain of the module and therefore the likelihood that later simultaneous presentation of the patterns will lead to module activity reaching a limit. Local pattern co-occurrence will form local pattern associative bonds, letting larger groupings act as a unit, that is, a unit that increases and decreases in activity together. Large-scale patterns will tend to bind many module activities together since learning takes place embedded in a larger informational structure. Such loops are reminiscent of the Bidirectional Associative Memory [BAM] of Kosko and will have similar dynamics [8].

2.5 Module Assemblies

If two modules are reciprocally associatively linked, we have a situation similar to the Bidirectional Associative Memory. If there are multiple interacting modules, we have the potential to form other interesting and complex associatively linked structures through Hebb learning, what we will call **module assemblies**, in harmony with the picture of visual information processing seen in primate inferotemporal cortex [9-11].

The source for this idea is the **cell assembly**, first proposed by Donald Hebb in his 1949 book *Organization of Behavior* [12]. What has become known as the **Hebb Rule** for synaptic coupling modification was originally proposed specifically to allow for the formation of cell assemblies.

A cell assembly is a closed chain of mutually self-exciting neurons. Hebb viewed the assembly as the link between cognition and neuroscience. When an assembly was active, it corresponded to a cognitive entity, for example, a concept or a word. Although the idea is an appealing one, it is hard to make it work in practice because it is difficult to form stable cell assemblies. Two common pathological situations are (a) no activity in the network after a period due to inadequate gain around the loop and, (b) spread of activity over the entire network since neurons in a realistic model system will participate in multiple assemblies and activity will spread widely. It is possible to control this behavior to some degree by making strong assumptions about inhibition, but the resulting systems are not robust.

As we mentioned, a key assumption of the Network of Networks model is that the basic computing elements are interacting groups of neurons. Module activity is not a scalar but a pattern of activity, that is, a high dimensional vector. Connections between modules are in the form of interactions between patterns. There is an intrinsic degree of selectivity. Patterns are less likely to spread promiscuously.

Because of this increased selectivity it might be possible that several nearby modules can become linked together to form loops through Hebb learning and can remain stable structures. We showed in Section 2.4 that associatively connecting modules together can increase the feedback coefficients in both modules.

2.6　Interference Patterns and Traveling Waves

Because we have suggested many important connections are local, much information processing takes place by movement of information laterally from module to module. This lateral information flow requires time and some important assumptions about the initial wiring of the modules. There is currently considerable experimental data supporting the idea of lateral information transfer in cerebral cortex over significant distances. The lateral information flow allows the potential for the formation of the feature combinations in the interference patterns, useful for pattern recognition. There is no particular reason to suggest that modules just passing information are in attractor states; for pattern transmission it is better if they are not.

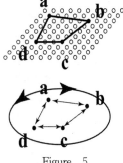

Coincidences, where two patterns collide are of special interest. Since the individual modules are nonlinear learning networks, we have here the potential for forming new attractor states when an interference pattern forms, that is, when two patterns arrive simultaneously at a module over different pathways.

Figure 5

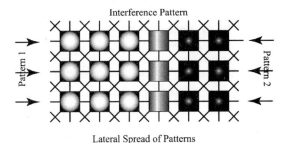

Interference Pattern

Pattern 1

Pattern 2

Lateral Spread of Patterns

Figure 6

There have been a number of related computational models specifically designed for vision that have assumed that image processing involves lateral spread of information. An early example is Pitts and McCulloch [13] who suggested, "A square in the visual field, as it moved in and out in successive constrictions and dilations in Area 17 (now called V1), would trace out four spokes radiating from a common center upon the recipient mosaic. This four-spoked form, not at all like a square, would be the size-invariant figure of a square (p. 55)." In the 1970's Blum[14] proposed the "grassfire" model where visual contours ignited metaphorical "grassfires" and where the flame fronts intersected produced a somewhat size invariant representation of an object. The propagating waves are computing something like the **medial axis representation**, that is, the point on the axis lying halfway between contours [15].

There are many examples of traveling waves in cortex. Bringuier, Chavane, Glaeser, and Fregnac [16] observed long range interactions in V1 with an inferred conduction velocity of approximately 0. 1 m/sec. Lee, Mumford, Romero, and Lamme[17] discussed units in visual cortex that seem to respond to the medial axis. Particularly pertinent in this context is Lee [18] who discussed medial axis representations in the light of the organization of V1. In psychophysics, Kovacs and Julesz[19] and Kovacs, Feher, and Julesz [20] demonstrated threshold enhancement at the center of circle and at the foci of ellipses composed of oriented Gabor patches forming a closed contour. These models assume that an unspecified form of "activation" is being spread whereas the Network of Networks assumes that pattern information (a vector) related to module attractor states is being propagated. We feel that the traveling wave mechanism and its generalizations may have more general applications that vision.

2.7　Ersatz Hardware: A Brief Sketch

How hard would it be to implement such a cortex-like system in hardware? This section is a "back of the envelope" estimate of the numbers. We feel that there is a size, connectivity, and computational power sweet spot about the level of the parameters of the network of network model. If we equate an elementary attractor network with 10^4 actual neurons, that network might display perhaps 50 well-defined attractor states. Each elementary network might connect to 50 others through 50×50 state connection matrices. Therefore a cortex-sized artificial system might consist of 10^6 elementary units with about 10^{11} to 10^{12} (0. 1 to 1 terabyte) total strengths involved to specify connections. Assumed each elementary unit has the processing power of a simple CPU. If we assume 100 to 1000 CPU's can be placed on a chip, there would be perhaps 1000 to 10,000 chips in a brain

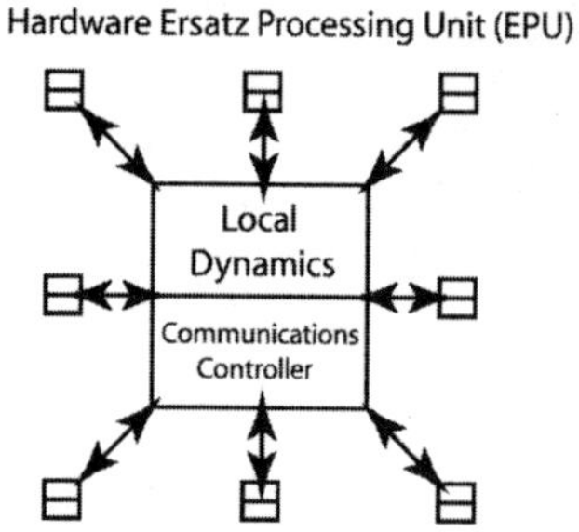

Hardware Ersatz Processing Unit (EPU)

Local Dynamics

Communications Controller

Figure　7

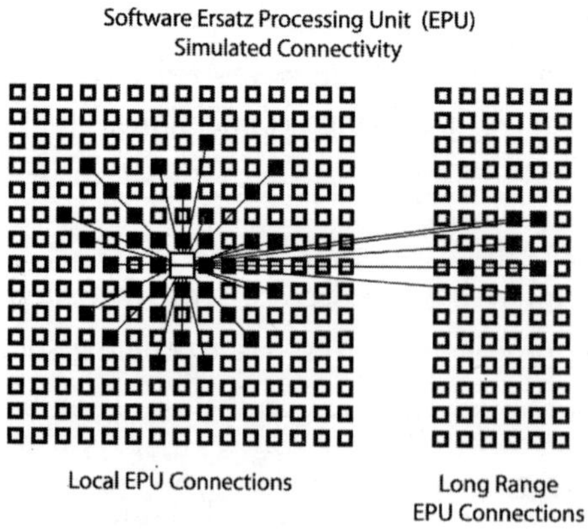

Software Ersatz Processing Unit (EPU)
Simulated Connectivity

Local EPU Connections　　　Long Range EPU Connections

Figure　8 .

sized system. These numbers are within the capability of current technology.

Therefore, our basic architecture consists of a large number of simple CPUs connected locally to each other and arranged in a two dimensional array. A 2 – D arrangement is simple, cheap to implement, and corresponds to the actual 2 – D anatomy of cerebral cortex. Intrinsic 2 – D topography can also make effective use of the spatial data representations used in cortex for vision, audition, skin senses and motor control.

2. 8   Communications

The brain has extensive local and long-range communications. The brain is unlike a traditional computer in that its program, dynamics, and computations are determined primarily by strengths of its connections. Details of these relatively **sparse interconnections** are critical to every aspect of brain function.

Perhaps as a consequence, the brain also displays **sparse representation** at its highest levels, that is, only a "few" units are active to represent a complex stimulus[9-11, 21].

2. 8. 1   Short-range connections

There is extensive local connectivity in cortex. An artificial system has many options. The simplest connections are purely to its neighbors. Expanding local connectivity to include modules two or three modules away, often seems to work better but is more complex to build.

2. 8. 2   Long-range connections

Many of the most important operations in brain computation involve pattern association where an input pattern is transformed to an associated output pattern that can be different from the input. One group of units connects to other groups with precisely targeted long-range connections.

2. 8. 3   CPU Functions

The CPUs must handle two quite different sets of operations. First, is communications with other CPUs. Much of the time and effort in brain-based computation is in getting the data to where it needs to be. Second, when the data arrives, it can then be used for detailed computation. The illustrations suggest some of the connections of the basic hardware module, the Ersatz Processing Unit, or EPU.

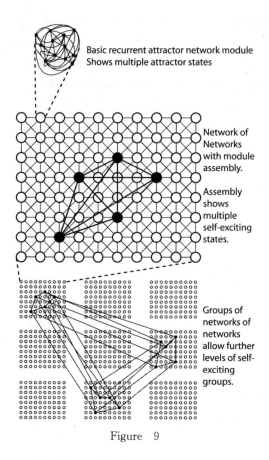

Figure 9

2. 8. 4 Topography: Bug or Feature

Moving information around the NofN Array can be slow since multiple steps in the EPU are involved. This might be considered a bug, but we view it as an Ersatz feature since the topographic arrangement of data on the array becomes a key component of the program and the computation. The cortex, and, in imitation, our computational model is a 2-D sheet of modules. Connections between modules, slow and expensive in the brain, as we have noted, perform the computation. Therefore topographic relationships between modules and their timing relationships become of critical computational significance. We suggest that it is possible to use the topographic structure of the array to perform some interesting computations. Indeed, **topographic computation** may be a major mechanism used to direct a computation in practice, and, therefore, a tool that can be used to program the array.

There are some interesting physical precursors to computation using an essential topographic aspect. We base many of our computational models on the arrangement and movement of data on a two dimensional sheet, the Network of Networks array. There are useful, even inspirational, examples of this architecture in optics and, most notably, in the design and function of Surface Acoustic wave filters, now ubiquitous in TV sets and cell phones [22].

2.9 Summary of Ersatz Brain Fundamentals

The Ersatz Brain is constructed from many small attractor networks (modules) sparsely connected into a Network of Networks. It is capable of performing specific cognitive tasks. It has both a discrete and continuous set of control structures capable of performthese specific tasks through explicit, teachable, cognitive programs.

Processing moves up in scale (Illustration) from

- Inflexible but highly optimized data sensors providing the inputs to modules.
- Through sparse module assemblies formed by learning in a single NofN array of modules.
- To flexible programs involving representations learning and dynamical systems working at multiple spatial scales. The hierarchical computation is fast, parallel, and highly integrative.
- As the joke has it, It's dynamical systems all the way down.

3 Ersatz Brain Software: Identity and Symmetry

Computation of identity and symmetry can serve as simple examples of what an "Ersatz" program might look like and the kinds of mechanisms required to make it work. Cognitive science suggests that brain-like computers will not be based on logic, but will be essentially alogical, that is, without logic, basing their operation on association, approximation, and analog operations. Some "mind" computational mechanisms include highly effective destructive data compression, storage of essence, as opposed to storage of detail, use of best approximators such as concepts, categories, and the use of analogy, metaphor and anecdote. Many, perhaps most, of these techniques are required because of the severely limited biological capacity of memory and the brain's weak ability to compute in the classic sense to implement human cognition.

We emphasized the importance of local computation and "traveling waves", that is, patterns of activity moving physically in cerebral cortex. Let us use these "traveling waves" and see what we can do with them. We observed that the waves travel slowly, perhaps 0.1 m/sec. Time plays an essential role.

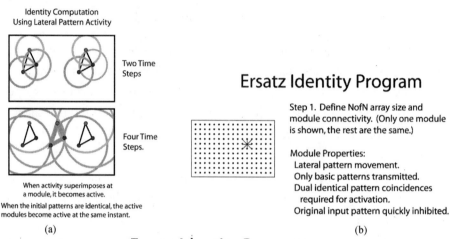

(a)　　　　　　　　　　　　　　　　　　　　　(b)

(c)

Figure　10

3.1　Identity

Let us look at a simple version of "Identity", a classic operation in computation. Consider two identical figures on a Network of Networks array as shown in the illustration. A figure here is defined as a set of several features in a particular physical arrangement. Consider traveling waves emanating from each feature comprising the two figures. At some point the two waves will collide. If the same feature (a pattern) is present in both figures, half way between them there will be a large burst of activity of that pattern at the collision point. The illustrations shows only the earliest collision; there will be others as the wavefronts collide. (We can use these later interactions for other purposes.)

Ersatz Identity Program

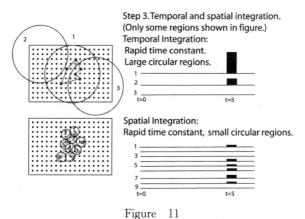

Step 3. Temporal and spatial integration.
(Only some regions shown in figure.)
Temporal Integration:
Rapid time constant.
Large circular regions.

Spatial Integration:
Rapid time constant, small circular regions.

Figure   11

Because the figures are identical by assumption, all their component features will collide half way between the figures, constructing a copy of the figure at the midpoint, a medial axis ghost. Notice that we are not really computing "identity" but "very highly similar". In fact, humans are quite bad at detecting small differences between very similar figures, an observation that leads to a whole series of popular puzzles for children and adults.

Ersatz Identity Program

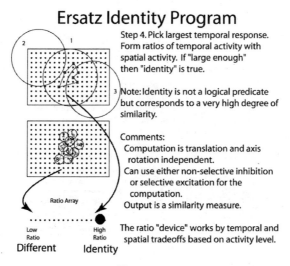

Step 4. Pick largest temporal response.
Form ratios of temporal activity with spatial activity. If "large enough" then "identity" is true.

Note: Identity is not a logical predicate but corresponds to a very high degree of similarity.

Comments:
Computation is translation and axis rotation independent.
Can use either non-selective inhibition or selective excitation for the computation.
Output is a similarity measure.

The ratio "device" works by temporal and spatial tradeoffs based on activity level.

Figure   12

We can now make a slightly more detailed version of this observation. The sequence of steps will be presented as a sequential series of "slides".

Notice that the operations in the third and fourth steps are based on the output of analog filters of various spatial and temporal extents.

Symmetrical Figures

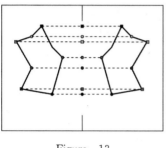

Figure   13

The key observation is that all the features from both figures collide simultaneously, giving rise to a large burst of integrated activity. However, this activity is spread out over a region of space, the size of the figures. Analog filters can be constructed to respond to this spatio-temporal pattern. Clearly effort will be required to define temporal and spatial filters in terms of time, space, required accuracy, and normalized figure size but simulations suggest it is not difficult to build filters providing an acceptable measure of "identicalness."

3.2   Symmetry

Symmetrical figures have the same features as identical figures, the same distances apart, but in a different spatial arrangement for symmetry reflected along a vertical line. In this case, the feature pattern collisions are all along the midline but occur at different times. A different set of analog

Ersatz Symmetry: Step 3

Figure   14

filters looking for temporal dispersion and spatial localization can detect symmetry. Again, designing the filter settings will take work, gains must be normalized, and so on, but we feel this can be done reliably, based on sound engineering principles.

Identity can be determined using this technique if the two figures are located at any position on the array. This is not true for symmetry where the resulting collision forms changes drastically with the position of the figures. This observation suggests that identity is a more salient perceptual property than symmetry, an observation that is probably correct for human perception.

4 Vowel Formant Invariance

We were interested in working with the Ersatz approach on a perceptual transformation that is commonly seen in speech and that may occur in other sensory applications as well. We looked at vowel data because good experimental data is available[23,24]. We are not speech scientists but this idea may be useful for other applications.

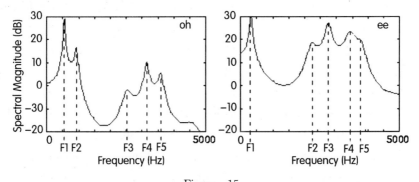

Figure 15

Vowels are long duration speech signals andare sometimes stable in their structure over a couple of hundred milliseconds. Speech signals are characterized by their "formants" which correspond to resonances of the vocal tract. The illustration shows examples from two vowels. The pattern of the resonances gives rise to the perception of vowels. (Other speech sounds can be more complex in their acoustics.)

However, there are serious problems dealing with vowels in a signal processing context. Among them:

• High variability of examples from a single speaker.

• Different speakers.

• Accents.

• Dipthongs (shifts in formant frequenciesas a vowel progresses).

• Context effects (neighboring phonemes change the resonances).

• Age.

• Gender.

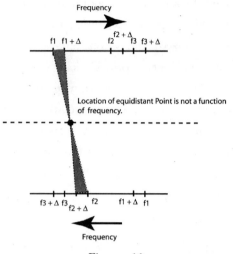

Figure   16

Resonances are a function of the size of the vocal tract, which differs among individuals, most notably between men, women, and children.

If the shape of the resonator does not change, then the resonances will scale with size. (Think organ pipes.)   So the resonances between different tract lengths will scale by multiplication by a constant.

Multiplication by a constant will mean the ratios of the formants will remain the same with different tract sizes.

Although the same formant can vary over 30% in frequency between men, women, and children, the ratios between them are far more stable, obviously reducing difficulties for pattern recognition.

This multiplicative scaling rule has an additional consequence. If we have a set of formant frequencies $\{f_i\}$ that can be scaled by a constant c, that is, $\{cf_i\}$ and if we take the log of this quantity it can be represented as log c $+ \{\log f_i\}$, that is, the log formant frequencies have been translated by a constant amount, log c. Since we know that frequency is represented on the $2-D$ surface of cortex with a roughly logarithmic map, the shift in formants corresponds to a shift of the pattern of log formant frequencies a constant distance, Δ, on the surface of cortex.

We know there are many frequency maps in early auditory cortex. There are

three core regions and perhaps a dozen
more in a "belt" surrounding the core.
We suggest we can use these maps to
compute formant ratios[25, 26].

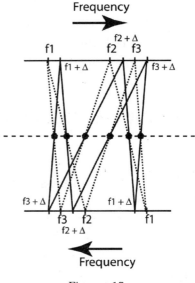

Figure   17

Our computational goal is en-
hance the ratios of the formants and
de-emphasize their absolute values.
We will use the lateral pattern spread
we used in the last section.

There is a simple geometry that
will let us do this. Consider two log
frequency maps running in opposite
directions. If each formant emits lateral waves, and a change in vocal
tract length corresponds to a constant
displacement, then there will be a line halfway between the two maps where
there is no effect of displacement.

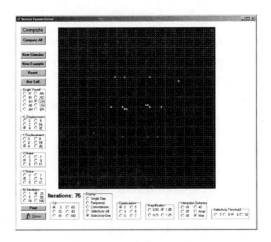

Figure   18

If we have multiple formants, each interacting, then we may have several invariant points along the midline that might uniquely represent each
vowel, irrespective of vocal tract length, which is what we want.

The screen shot displays the white invariant points for a vowel in the
middle between the two opposite sense formant log frequency components at

the top and bottom.

As an extra, speculatively, the same geometrical trick would let us represent in the same way with spatial representations of two orthogonal parameters, for example, frequency and amplitude.

5　Potential Extensions to Higher Level Cognition

5.1　Active and passive memory

Many cognitive scientists believe we have at least two systems of reasoning. One is slow, deliberate, used logic and reasoning. It is what we are taught to do in school though most humans can be shown to be very bad at it. But there is another "reasoning" system that is basically associative, alogical, intuitive, very fast and right much of the time. It is this second system, frequently used successfully by humans in daily life that we wish to model with the Ersatz Brain. This dichotomy is sometimes called "two process theory" in cognitive science [27].

Malcolm Gladwell's recent best seller, Blink, made this point clearly:

"We believe that we are always better off gathering as much information as possible and spending as much time as possible in deliberations. There are moments ···when our snap judgments and first impressions can offer a much better means of making sense of the world. The first task of Blink is to convince you of a simple fact: decisions made very quickly can be every bit as good as decisions made cautiously and deliberately. " [28]

Simple Associative Link

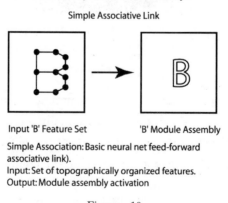

Input 'B' Feature Set　　　　　'B' Module Assembly

Simple Association: Basic neural net feed-forward associative link).
Input: Set of topographically organized features.
Output: Module assembly activation

Figure　19

Most traditional machine learning works with a very special problem that is basically pattern recognition. We have a set of known categories. We have a set of input data. The idea is to learn to use the input data so as to correctly classify new inputs in the right categories with some degree of precision. Variants include unsupervised learning and clustering algorithms, but the idea is generally to learn to categorize input data in some natural way.

Such systems are useful in a number of specific practical applications, for example, character recognition. However human use of cognitive operations is much more flexible and task oriented. It is also very fast. These virtues make up in practice for its many errors.

Next we will consider extensions of the Ersatz ideas we sketched in the earlier parts of this paper, based on experiments and insights from cognitive science.

5. 2 Passive Memory

The first class of mental operations we shall call **passive** memory. It roughly comprises what is called "machine learning". There is a memory representation consisting of a number of categories, for example, letters or digits. There is input data that corresponds to one or another member of the category. The feature set leads to activation of a category that will be correct to some degree of certainty. This process is well defined, mechanical, but is not flexible. It is diagrammed in the illustration.

5. 3 Active Memory

Let us suggest an extension of this idea we shall call active memory that is dependent on the particular task to be performed. Consider the visual pattern "B". The input features of this visual pattern are constant. For humans, however, depending on the specific task, this pattern could be categorized as 'letter B', a 'capital letter B', a 'black capital letter B', 'an alternative Plan B', 'a black Courier capital letter B', or 'a good exam grade'. All these interpretations are correct answers for one specific cognitive task. For some of these tasks the different pattern, lower case 'b' would be in the same class; for the other tasks it would not. In applications in practice, there can be no further learning, that is, the input pattern can give rise to several possible answers based on re-arrangement of past knowledge. That is, the combination of specific task and previous learning requires no training period. (Example: Is 73 larger than 88?) Therefore for practical cognitive applications it will be necessary to specify the task properly in addition to learning the specific patterns involved at some past time. Worse, the categorizations can change after the pattern and its associations are learned. The categories are determined by the task as much as learning input data with powerful machine learning algorithms. Attempts to store the task as part of the data have not worked well, in general.

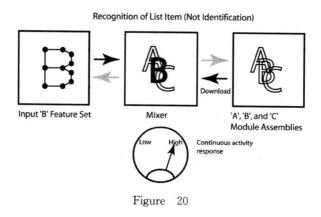

Figure   20

5.4   Implementing Active Memory

Traditional computers contain an essential register or registers called the **accumulator**. Accumulators are where information interacts with other information. We suggest the usefulness of what we will call **a mixer**, where information from other regions or the task itself comes to interact with input information to accomplish a specific task, as shown in the illustration.

Our mixer has (at least) two modes of interest to us.

First, as indicated by the meter face, it decides tasks where the output can be true, false or a scalar value, for example, a strength measure.

Second, it can integrate its multiple inputs to form a new output pattern.

Intermediate outputs — the relative strength of a pattern — are possible as well.

5.5   A Test Problem for Active Learning: The Sternberg Task

Consider a simple test problem. This problem is based on what in cognitive psychology is called the "Sternberg paradigm" after a well-known set of experiments by Saul Sternberg, at that time at AT&T Bell Labs[29].

These experiments provided the first strong evidence that many mental operations had a major parallel component.

This experiment created a great deal of interest and innumerable variations were studied.

The experiment is simple. Suppose we learn a list of items taken from a larger set, for example, letters, numbers, drawings, etc. The task of the subject is to decide whether a presented item is present in the learned list.

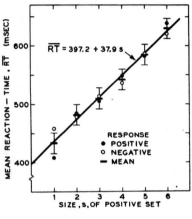

Figure   21

As an example, suppose the subject learned a list of letters, "B,C,U,X". If the letter 'C' is presented, the subject makes a positive response. If the letter 'A' is presented, the subject makes a negative response. If there was a list of separate items in memory, on the average of half the list would be scanned for a match but the entire list should be scanned if the item was not on the list. As the experimenter increased the number of items on the list, response time for negative responses should have twice the slope of positive items. But this is not seen. Both positive and negative items show the same slope, which seems inefficient unless for some reason an "exhaustive scan" of the list is required.

Parallelism, of course, can naturally implement an exhaustive scan while serial operations do not do so. The illustration shows data from Sternberg. The x-axis gives the number of items on the list and the y-axis the response time. Note that negative (not on the list) and positive (on the list) response times have the same slope, evidence for an "exhaustive scan" of the list. This experimental result holds for a large number of list item types —

letters, numbers, faces, unfamiliar shapes.

In one of many variants of these experiments, the subject is shown a new list for every test. Subjects in this task seem to develop immediately an **attentional filter** of some kind based on the guided manipulation of previously learned material so as to let them perform the task. An approach very similar to that presented next, and on which the current model is based, can be found in [30].

Use of our ideas about mixers suggests a simple way to compute on this problem.

We can program this attentional operation by exciting the list item patterns in the memory. The set of excited representations (in Ersatz language, module assemblies) projects to the mixer. The test item also projects to the mixer. If there is a match of a component of the list, there is a strong positive analog response of the mixer (symbolized by the meter pointer on the mixer). If none of the items is identical, there is a slower negative response. Note the simultaneous presence of active representations of all the list items at the mixer, letting the computation be done in parallel. The mixer contains the sum of activated list items, which may interfere with each other. It is possible to calculate the signal to noise ratio of the responses for this task and show it is adequate, in fact, optimal, for implementing recognition accuracy.

5.6　A Succinct Program Description

We can make this "program" a little more formal.

Initial Conditions: For this example we assume 26 letters represent the categories in the stimulus set.

Module Arrays: Three NofN **Arrays, Array A, Array B,** and **Array C.**

Connections: In line, reciprocal

$$A \Leftrightarrow B \Leftrightarrow C$$

Array A: Sensory Input **Array** $(a, b, c, \cdots z)$

Array B: Intermediate representations

　　Assemblies $\{a, b, c \cdots z\}$

Array C: Classifications (Stable Module Assemblies)

　　Assemblies: $\{A, B, C \cdots, Z\}$

Associations are formed through earlier learning:

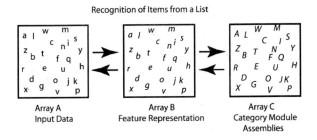

Figure 22

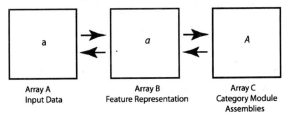

Figure 23

Array A ⟺ Array B ⟺ Array C

Learned Data Patterns

ay $\Rightarrow$ *a* $\Rightarrow$ ***A***

bee $\Rightarrow$ *b* $\Rightarrow$ ***B***

cee $\Rightarrow$ *c* $\Rightarrow$ ***C***

...

zee $\Rightarrow$ *z* $\Rightarrow$ ***Z***

When the categories are forming, we are also learning the reciprocal associations:

ay $\Leftarrow$ *a* $\Leftarrow$ ***A***

bee $\Leftarrow$ *b* $\Leftarrow$ ***B***

cee $\Leftarrow$ *c* $\Leftarrow$ ***C***

...

zee $\Leftarrow$ *z* $\Leftarrow$ ***Z***

Operating Program.

List Formation. A **list** is a subset of the number of categories (module

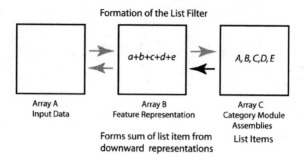

Figure　24

assemblies) in **Array C,** for example, $\{A,B,C,D,E\}$.

Present the list in sequence at **Array C,** to form summed associative representations in **Array B.**

Step 1. Form list representation by sequential presentation:

Array A　　　Array B $\Leftarrow$ **Array C**

$$a \Leftarrow A$$
$$a+b \Leftarrow B$$
$$a+b+c \Leftarrow C$$
$$a+b+c+d \Leftarrow D$$
$$a+b+c+d+e \Leftarrow E$$

Step 2. Interaction

Pattern 'a' is input.

Array A　　　Array B　　　Array C

$$a \Rightarrow a+b+c+d+e$$

Filter output: $[a,a] + noise$

That is, when input and memory are simultaneously active, we form the inner product between summed list in **Array B** and the input pattern, here 'a'.

If a pattern is present, there will be a large response of the list filter. A nice aspect of this approach is that the list filter can be shown to implement the optimal linear filter for the detection of a known signal in noise, the matched filter, that is, this strange cognitive model is also optimal for the detection task.

Step 3. Decision Criteria (very tentative):

Positive: **Large, rapid, synchronized compact activation pattern in** Array

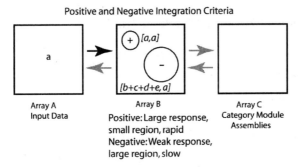

Positive and Negative Integration Criteria

Array A
Input Data

Array B
Positive: Large response,
small region, rapid
Negative: Weak response,
large region, slow

Array C
Category Module
Assemblies

Figure   25

B **is** positive.

Negative; **Smaller, slow, unsynchronized, widespread activation pattern
in** Array B **is** negative.

The decision parameters are set by the accuracy required, the nature of
the categories, etc. Previous task and component experience is required.

An Unintuitive Prediction and Basic Level Categories.

Note the unintuitive prediction of this idea that the interaction for presence of the item on the list does not require identifying the list item, that is,
the system does not reason, "I see a C. ⇒ C is on the list. ⇒ Therefore I
will make a positive response. " The process seems to be "The item is some
item on the list, one that gives a large response from the list filter". ⇒
Therefore I make a positive response. ⇒ And with a little more processing
effort (taking time) I conclude the item is, in addition, a C. " This curious
prediction seems to be experimentally correct.

This example points out one important aspect of a practical cognitive
system. It makes sense to operate at a level of generality well above the
finest detail, even think the fine detail is available and one might think that
use of more features would make for faster and more accurate discriminations.

Human categories, presumably the result of millennia of learning about
the real world, by default operate at what is called the "basic level" with just
enough detail to be useful but not so much as to overwhelm the cognitive apparatus with excess data. A "table" can be large and mahogany or small and
plastic, but if the task is only to differentiate tables from chairs because you

want to sit down, there is no need to make finer discriminations[31].

5.7　Generalization and Use of the Mixer.

We are now in a position to suggest a more general computational system based on the active memory. Initially, we assume the programming is done by atask controller located "somewhere else".

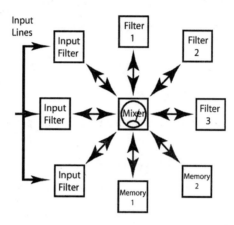

Figure　26

Each region in the Ersatz Array connects to multiple other regions. Each region sets up its own filters, based on its predetermined computational wiring. Partial results are available quickly, though at the price of less accuracy, another characteristic of human cognition with substantial experimental support.

The illustration shows the general architecture of the Ersatz cognitive computing system.

- task driven;
- memory based;
- attentionally programmed;
- multimodal: all senses project to the mixer;
- non binary tasks can use pattern outputs.

We are close to implementing this architecture as part of the Ersatz Programming Environment.

One of the Project's interim goals is to join the two development pathways, Active and Passive memory, so that problems in cognitive signal processing can be quickly, easily, and flexibly programmed by a user and so that the resulting Ersatz Array code will run quickly and efficiently.

5. 8   Context

Real world problems can take a while to describe. Very little linguistic behavior consists of single isolated sentences. Real linguistic behavior consists of multiple sentences on a single topic. There are a number of ways of developing context.

We had some past success using hierarchical networks to represent knowledge for tasks like disambiguation. Let us use simple "neighborhood" association rather than a structured net. Then we can suggest some natural models for interesting cognitive tasks like disambiguation. These are probably not the best models, but they might be a place to start.

There are at least two kinds of superposition of states that are relevant. First, in a single module, the linear region of the module is used for the weighted superposition of multiple attractor states. Second, module assemblies are sparsely coded. Therefore two different module assemblies are unlikely to have many modules in common. Therefore in a region of the Ersatz array, multiple states can be superimposed.

We can use this observation make a very simple, though sloppy, context neighborhood that looks a little like a semantic network by assuming that previous input words hang around and are associatively linked to other words. Perhaps an entire semantic network could be constructed from the union of neighborhoods, but this may not really be necessary. Assume that word patterns from multiple sentences superimpose.

As a simple-minded example, let us consider a set of four sentences on a common topic.

I needed caffeine.

I walked to Starbucks.

"A vente ," I said to the barrista.

"Ah, java", I exclaimed.

What is the appropriate meaning of "java" in this sequence? The string "java" can be a slang term for coffee, an island in Indonesia, or a programming language.

In this case, superposition of the associations of some of the words in the previous sentences gives rise to a context that easily disambiguates

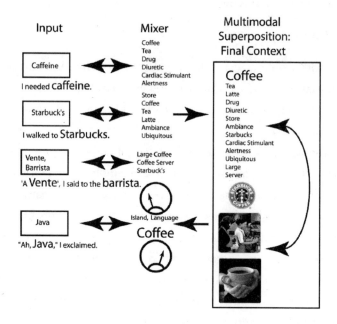

Figure　27

"java" in the last sentence as suggested in the illustration.

6　Conclusions

This brief discussion of the Ersatz Brain organization has tried to suggest that it is a practical, scalable, flexible, and powerful architecture for computing machines that need to perform efficiently certain classes of tasks that are often thought to be typical of human cognition.

Acknowledgement

The authors would like to acknowledge support for this project from the AFRL (Air Force Research Laboratory, Rome, NY and currently DARPA (Defense Advanced Research Projects Agency, Washington, DC).

References

[1] Wikipedia entry on "Artificial Intelligence", May, 2011

[2] Wikipedia entry on "Computer Chess", May, 2011

[3] Wikipedia entry on "Watson (computer)", May, 2011

[4] Anderson JA, Sutton JP. If We Compute Faster, Do We Understand Better? *Behavior Research Methods, Instruments, and Computers*, 1997(29): 67 – 77

[5] Anderson JA. Allopenna P, Guralnik GS, *et al.* Programming a Parallel Computer The Ersatz Brain Project. In: Duch W, Mandzuik J, and Zurada JM(eds.). *Challenges to Computational Intelligence*. Springer, Berlin (2007)

[6] Mountcastle VB. Introduction. *Cerebral Cortex*. 2003(13): 2 – 4

[7] Anderson JA. *The BSB Network*. In: Hassoun MH(ed.); *Associative Neural Networks*. Oxford University Press, New York (1993):77 – 103

[8] Kosko B. Bidirectional Associative Memory. *IEEE Transactions on Systems, Man, and Cybernetics*, 1988(18):49 – 60

[9] Tsunoda K, Yamane Y, Nishizaki M, *et al.* Objects are Represented in Macaque Inferotemporal Cortex by the Combination of Feature Columns. *Nature Neuroscience*, 2003(4): 832 – 838

[10] Tanaka K. Inferotemporal Cortex and Object Vision. In: Cowan WM, Shooter EM, Stevens CF, Thompson RF(eds.). *Annual Review of Neuroscience*, 1996(19): 109 – 139

[11] Tanaka K. Columns for Complex Visual Object Features in Inferotemporal Cortex: Clustering of cells with similar but slightly different stimulus selectivities. *Cerebral Cortex*, 2003(13): 90 – 99

[12] Hebb DO. *The Organization of Behavior*. Wiley, New York (1949)

[13] Pitts W, McCulloch WS. How We Know Universals: The Perception of Auditory and Visual Forms. In: McCulloch WS (ed., 1965): *Embodiments of Mind*. MIT Press, Cambridge, MA (1947/1965) 46 – 66

[14] Blum HJ. Biological Shape and Visual Science (Part I). *Journal of Theoretical Biology*, 1973(38): 205 – 287

[15] Kimia B, Tannenbaum A, Zucker SW. Shapes, Shocks and Deformations {I}: The Components of Shape and the Reaction-Diffusion Space. *International Journal of Computer Vision*, 1995 (15):189－224

[16] Bringuier V, Chavane F, Glaeser L, Fregnac Y. Horizontal Propagation of Visual Activity in the Synaptic Integration Field of Area 17 Neurons. *Science*, 1999(283): 695－699

[17] Lee T－S, Mumford D, Romero R, Lamme VAF. The Role of Primary Visual Cortex in Higher Level Vision. *Vision Research*, 1998(38): 2429－2454

[18] Lee T－S. Analysis and Synthesis of Visual Images in the Brain. In: Olver P, Tannenbaum A (eds.). *Image Analysis and the Brain*. Springer, Berlin, 2002: 87－106

[19] Kovacs I, Julesz B. Perceptual Sensitivity Maps Within Globally Defined Shapes. *Nature*, 1994(370): 644－646

[20] Kovacs I, Feher A, Julesz B. Medial Point Description of Shape: A Representation for Action Coding and its Psychophysical Correlates. *Vision Research*, 1998(38): 2323－2333

[21] Olshausen BA, Field DJ. Sparse Coding of Sensor Inputs. *Current Opinions in Neurobiology*, 2004(14): 481－487

[22] Morgan DP. *Surface Acoustic Wave Filters with Applications to Electronic Communication and Signal Processing* (2nd Ed.). Academic Press, New York (2007).

[23] Watrous RL. Current Status of Peterson-Barney Vowel Formant Data. *Journal of the Acoustical Society of America*, 1991(89): 2459－2460

[24] Peterson GE, Barney HL. Control Methods Used in a Study of the Vowels. *Journal of the Acoustical Society of America*, 1952 (24):175－184

[25] Petkov CI, Kayser C, Augath M, Logothetis NK. Functional Imaging Reveals Numerous Fields in the Monkey Auditory Cortex. (2006) *PLoS Biol* 4 (7): *e215. doi: 10. 1371/journal. pbio. 0040215*

[26] Talavage TM, Sereno MI, Melcher JR, *et al*. Tonotopic Organization in Human Auditory Cortex Revealed by Progressions of

Frequency Sensitivity. *Journal of Neurophysiology*. 2004(91):
1292 – 1296

[27] Sloman S. *Causal Models: How People Think About the World and Its Alternatives*. Oxford University Press, New York, NY (2005)

[28] Gladwell M. *Blink*, Little, Brown, Boston, MA (2005),13 – 14

[29] Sternberg S. High speed scanning in human memory. *Science*, 1996(153), 652 – 654

[30] Anderson JA. A theory for the recognition of items from short memorized lists. *Psychological Review*, 1973(80): 417 – 438.

[31] Murphy GL. *The Big Book of Concepts*. MIT Press, Cambridge, MA (2002)

讲座人简介

James A. Anderson is now a Professor in the Department of Cognitive, Linguistic and Psychological Sciences. He was Chair of the Department of Cognitive and Linguistic Sciences from 1993 to 1998 and in 2000−2001.

He has published extensively in the area of computational models for cognition and memory and computational neuroscience. He is currently working with colleagues from industry and Brown on the "Ersatz Brain Project" a design for a massively parallel "brain-like" computer.

Dr. Anderson is the author of many books and journal articles. Books include "Introduction to Neural Networks", "Neurocomputing", Volumes 1 and 2 and "Talking Nets: An Oral History of Neural Network Research", all from MIT Press,Cambridge, MA.

Dr. Anderson has a B. S. in physics (1962) and a Ph. D. in physiology (1967), both from the Massachusetts Institute of Technology. He has had postdoctoral fellowships at UCLA (1967−1971) and Rockefeller University (1971−1973), as well as a senior fellowship at the University of California, San Diego. (1979)

智能车辆的视觉认知计算

薛建儒

(西安交通大学,人工智能机器人研究所,西安,710049,中国)

摘要:视觉智能计算是无人车开发中最重要的问题。西安交通大学人工智能机器人研究所在 2003 年开始进行视觉和听觉智能计算研究,目的是开发能适应各种环境的智能车辆。我们提出了基于认知模型的智能驾驶系统框架,该框架建立在智能体的控制基础之上,将系统按感知决策和控制等功能进行分解,将多传感感知与融合计算和控制计算分离,可减少系统计算负担,并提高系统可靠性。针对驾驶场景中的道路和障碍物检测,该系统使用二级与或图表示视觉知识,使用显著图计算可能感兴趣的图像区域,建立了深度信息数据库用来计算障碍物的深度。在上述工作基础上,可对人类驾驶行为进行分析和建模,模仿人类的驾驶行为实现对无人车辆的智能控制。

关键词:智能车辆;视觉感知

Visual Cognitive Computing for Intelligent Vehicle

Jianru Xue

(Institute of Artificial Intelligence and Robotics, Xi'an Jiaotong University)

Abstract: Visual cognitive computing is an important problem in intelligent vehicle development. Institute of Artificial Intelligence and Robotics, Xi'an Jiaotong University started research on visual and audio cognitive computing in 2003. This project is funded by National Science Foundation of China. The aim of this project is develop intelligent vehicles that are suitable for various type of environment. Institute of Artificial Intelligence and Robotics, Xi'an Jiaotong

University proposed cognitive model based driving control framework，which is based on control of intelligent agent. The driving system is divided into perception decision and control. Synthesis computing of perceptions from multi-sensors and control computing are separated to reduce computation load and increase reliability. For road and obstacle detection in driving scene，visual knowledge is indicated by a two-level and-or-graph，and saliency map is used to calculate region of interest of the input image. They construct a depth information database to estimate the depth of obstacles. Human driving behavior is analyzed and modeled to control the unmanned vehicle.

Keywords：Intelligent vehicle，Visual cognitive

1　引言

视觉智能计算是无人车开发所面临的最重要的问题之一。在人类通过感觉器官获取的信息中，约有 75% 来自于视觉信息。现阶段大部分的无人智能车仍然依靠传统的计算模型，视觉感知在控制中所占比例并不大。因此，视觉智能计算将是未来智能车辆控制技术的发展方向。

图 1　复杂的驾驶场景

由于我们到现在为止还没能够完全了解人类视觉信息处理的机制，使得在此基础上建立的计算机对世界的感知和人类相比，无论从原理还是准确性上都存在较大的差距。

因此，针对智能车辆设计这个目标，我们尤其需要关注以下四方面的问题：

- 传感器数据获取和组织；
- 特征提取和选择；
- 物体识别和跟踪；
- 目标动作分析和行为识别。

根据上述内容,我们可将智能车辆中的视觉认知计算问题定义如下(图2):

- 多模态传感器数据处理 —— 视觉知识;
- 多传感器融合的驾驶场景感知 ——选择性关注(Selective attention)和反馈;
- 人类驾驶技术和行为的自主学习—— 人类技能的机器学习。

将上述三方面内容相结合,我们便可实现对无人车辆的控制。

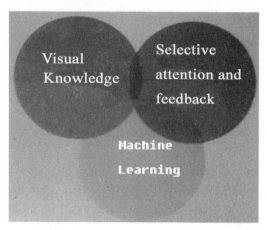

图2　视觉知识、选择性关注和机器学习三者的融合

2　驾驶场景中的视觉感知

人在驾驶时头部会转动,眼睛则会自动聚焦到感兴趣的深度内容。而智能车辆只能获得关于驾驶场景的二维图像投影,该驾驶场景包括了道路结构、交通信号、障碍物三个属性,形成了驾驶场景的几何结构。如何根据驾驶场景的几何结构从二维图像中获取有用的控制信息,这是一个需要研究的问题。

对于一幅输入图像,我们首先截取其中感兴趣的部分,利用视觉知识,将其分解为物体。视觉知识可用二级与或图(two-level and-or-graph)表示,并根据这些信息计算驾驶场景的属性,如图3所示。图4则是我们视觉感知系统的结构图。

为了丰富信息来源,我们也可使用多种传感器获取数据,包括视频、二维激光雷达、三维微波雷达等,通过将多传感器的数据进行融合,可获得驾驶场景的几何信息,并根据这些几何信息最终计算出道路的结构。

视觉感知系统的第二部分内容是选择性关注。人总是能把目光聚焦到自己感兴趣图像区域,大脑在视觉处理的各个阶段可自动提取感兴趣的内容,并丢弃大量不相关信息。因此我们开发了一个通用的关注选择系统,如图5所示,系统中有两条信息流，实线箭头表示主要信息流，虚线箭头表示模块影响。

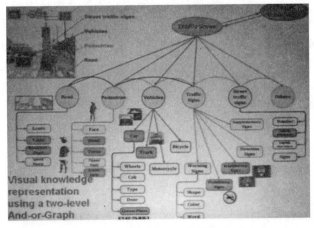

图3　二级与或图表示的视觉知识

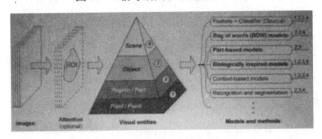

图4　视觉感知系统结构图

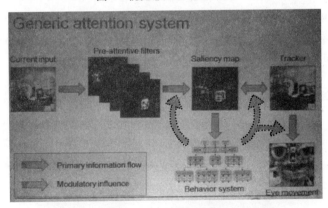

图5　通用关注选择系统

　　该系统是基于 Itti 等人提出的显著图（saliency map）方法。这种方法利用了神经生理学的研究成果，模拟人类的视觉注意力机制，认为图像中有强烈对比的部分会吸引人类的主要注意力，因此他们把像素与其背景的差异度定义为该点的显著值。结合显著值，可由颜色、亮度、方向等多种特征综合得到显著图,再根据显著图确定图像的显著区域。

因此这里的关键问题是如何计算显著图。在输入图像中,我们可以找到足够多的特征构建显著图。这些特征大体可分为两类,自下而上和自上而下。自下而上指的是从图像的角度出发挖掘出的特征,例如环绕中心的特征、尺寸无关的特征等。自上而下指的是从概念的角度出发挖掘出的特征,例如上下文特征、任务主导的特征等。我们综合利用这些特征计算图像的显著图,然后再根据显著图确定图像的显著区域。

3　障碍物的深度识别

人类驾驶行为严重依赖于障碍物的深度和层次识别。为了使得计算机能够进行深度识别,我们设计了深度和层次语义。例如,可设计车辆不同朝向时的层次化结构图,以及人体不同朝向的结构图。利用朝向层次结构图,我们可判断车辆和人的朝向、走向。

基于这种方法,又进一步设计了深度信息数据库。数据库中包括了人体、车辆、道路标志等驾驶场景中可能出现的物体深度信息标签。

给物体打上标签,一个难点问题是如何保证标签的正确性以及层次性。由于只用一个人进行判断结果风险较大,综合多人判断结果更符合统计意义,因此我们的关键问题是如何融合不同人给出的标签结果。投票转移(vote transferring)方法是一种有效的标签融合方法,可消除大部分的不一致结果。

图 6 是我们构造的一个简单有效的学习模型,所使用的学习算法可以在有限时间和空间内找到一个近似最优解。

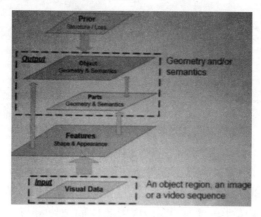

图 6　简单有效的学习模型

4　驾驶行为建模和控制

真实的驾驶场景往往十分复杂。一方面,全面获取相关道路信息有时是一件相当困难的事,通常无法确定道路拐角被建筑物遮挡的方向是否有车辆驶来,另一方面,针对具体的道路环境作出正确的决策也是一件有相当难度的事情,一旦发现道路上存在障碍,人可以很容易判断是该采取换道、转弯还是直线刹车等措施,而对于实验中的全自动智能车辆无法很好地进行处理。

因此,如何从人类驾驶员的驾驶行为中提炼出智能汽车可以理解并执行的驾驶能力是近年来持续的研究热点。目前最多的尝试是基于知识的驾驶行为学习,即研究如何将实际世界中的交通路况识别转化为知识世界中的特定场景,并激发相应的行为规则进行处理。然而,很多驾驶技能的学习并不能简单地用事件驱动的知识-规则系统分析,例如在车流行进中换道、在拥挤的停车场停车等就需要结合车辆的运动控制进行考虑。

由于复杂的车辆运动同时包含纵/横两个方向的运动控制要求,简单的解耦控制可能达不到相应的控制要求,对于这类情况下的车辆控制问题通常采用运动轨迹规划方法,一般可根据当前的车辆运行环境生成的合理轨迹,预测在目前的车流状态下,该驾驶员是否有时间进行换道/拐弯,或者引导驾驶员沿着合理的轨迹停好车。

此外,构造车辆轨迹产生系统时通常还可利用模糊系统和神经网络系统良好的非线性映射能力,例如根据已有的停车数据训练好的神经网络,可以以车辆的始、终点坐标和朝向作为输入,直接产生合适的转向控制时间序列和预期的运动轨迹。

5　总结

视觉智能计算是无人车开发中最重要的问题。现阶段大部分的无人智能车仍然依靠传统的计算模型,视觉感知在控制中所占比例不大。但视觉智能计算无疑将是未来智能车控制技术的发展方向。

西安交通大学人工智能机器人研究利用视觉感知计算,模拟人类的驾驶行为,设计出了无人驾驶机动车的原型系统,该项目得到了中国国家自然科学资金的资助。

参考文献

[1] Jianru Xue, Zheng Ma, Nanning Zheng. Hierarchical Model for Joint Detection and Tracking of Multi-target. *10th Asian Conference on Computer Vision 2009*

[2] Jianru Xue, Nanning Zheng, Xiaopin Zhong. Visual Perceptual Stimulus—A Bayesian-based Integration of Multi-Visual-Cue Approach and Its Application. *China Science Bulletin*, 2008, 53(2): 172-182

[3] Jianru Xue, Nanning Zheng, Jason Geng, Xiaopin Zhong. Tracking Multiple Visual Targets via Particle-based Belief Propagation. *IEEE Transactions on System*, *Man*, *and Cybernetics Part B*, 2008, 38(1): 196-209

[4] Jianru Xue, Nanning Zheng, Xiaopin Zhong. Sequential Stratified Sampling Belief Propagation for Multiple Targets Tracking. *Science in China*, *Issue. F*, 2006, 49(1): 48-62

[5] Itti L, Koch C, and Niebur E. A model of saliency-based visual attention for rapid scene analysis. Pattern Analysis and Machine Intelligence, IEEE Transactions on, 1998, 20: 1254-1259

[6] 李力,王跃飞,郑南宁,张毅. 驾驶行为智能分析的研究与发展. 自动化学报, 2007, 33:1014-1022

讲座人简介

薛建儒教授 2003 年毕业于西安交通大学,博士,教授,博士生导师,入选教育部新世纪优秀人才,获陕西省青年科技奖、陕西省青年突击手等奖项。研究方向为计算机视觉与模式识别;图像/视频处理与编码技术;机器学习与智能计算。

薛建儒教授作为主持人或主要研究者承担国家重点基础研究(973)、国家自然科学基金重点、国家重点科技攻关、国家"863"计划、国防基础预研、国际合作等科研项目 30 余项。目前主持和承担的研科研项目有:

(1) 主持国家自然科学基金项目——视频解析编码的随机图感知计算模

型与统计学习技术；

（2）主持国家"863"专项—— 新一代自适应数字视频编解码及传输关键技术研究与系统实现；

（3）主持科技部支撑计划项目——数字媒体内容互用技术与平台；

（4）副组长,国家自然科学基金重点项目——高度可伸缩数字视频编解码的基础理论和方法研究；

（5）副组长,国防基础预研项目。

薛建儒教授获国家发明二等奖 1 项(排名第四)，合作出版英文专著 1 部。发表国内外期刊和国际会议论文 30 余篇,其中国际期刊 4 篇,国际会议 20 余篇。获发明专利 7 项,获软件著作权 2 项。近 5 年国际学术会议特邀报告 3 次,组织和参与组织国际会议 7 次。

通往认知脑机接口

Jose C. Principe

(Electrical and Biomedical Engineering, University of Florida,
Gainesville, 32611 USA)

摘要:脑机接口(BMIs)的研究发展正面临着巨大的挑战,这里我们将从工程的角度总结脑机接口系统的几种不同的设计方法。佛罗里达大学计算神经工程实验室致力于运动皮层和伏核(Nucleus Accumbens)神经信号信息的提取,以及脑机接口系统中动作评价体系的建立。

关键词:脑机接口;动作评价;强化学习

Toward Cognitive Brain Machine Interfaces

Jose C. Principe

(Electrical and Biomedical Engineering, University of Florida,
Gainesville, 32611 USA)

Abstract: This talk will discuss current and future challenges of designing brain machine interfaces (BMIs). We will review the different designs developed for BMIs from an engineering perspective, and will present our current work on extracting information from the Nucleus Accumbens and Motor cortex to implement and Actor Critic Architecture for BMIs.

Keywords: BMI, Actor critic, Reinforcement learning

1 引言

让我们用一个有趣的开头来展示脑机接口研究的重要性。1990 年,我写了一份建议书,建议 CNEL 投资一项利用大脑活动性(EEG)的新的计算机接口技术(图 1)。那是 21 年前了,我们使用的是 NeXT 计算机,其优势在于有两个处理器(包含一个 DSP 处理器)。我的想法是利用事件相关电位(ERPs)来操作电脑。实验任务非常简单,将屏幕上的球从左边移到右边。计算机屏幕会不断闪烁"左"或"右"的字样,受试者就会根据自己的意愿想"是"或"不是"。DSP 处理器上实现了一个神经网络算法,实时地对单次事件相关电位进行分类。实验中,大约 2/3 的人可以成功完成实验,成功率高的可以达到 90%,但是一些人一次也没有成功过。进一步观察实验结果,系统的数据传输率约在 4.5 比特/分。这个速度相比于真实打印机慢很多,但对于那些有运动功能障碍的人来说,则是好处多多。重要的是,我们发现了脑机接口研究的可能性和用处。

图 1　一个简单的脑机接口系统

目前,大部分的脑机接口系统都是基于监督学习的框架建立的,即利用训练集中的数据训练解码算法,再用固定下来的解码算法解码剩下的数据。这种做法简单易行,但存在问题。我相信现在是时候跳出监督学习的框架来设计脑机接口系统。因为根据训练集训练出的脑机接口系统,纯粹只是一个大脑活动的传声机或翻译器,对于训练集之外的动作任务则无能为力。一种替代方法就是非监督学习(强化学习)。大脑是一个信号处理系统,可以从信号中学习外部世界的结构和信息;同时大脑是一个生理系统,只能从过去和现在的信息中学习,来预测未来的动作。因此,脑机接口系统应该具有类似于生物系统的学习能力,即不需要人特意告诉系统应该做什么,而由系统自己决定学习的时间和内容。我们把这种自适应的脑机接口系统称为共生脑机接口系统(Symbiotic BMI)。

实现共生脑机接口系统的方法有很多种,这里我们尝试提出两种系统框架:一种是基于奖赏的脑机接口系统,另一种是目标驱动的脑机接口系统。

2 基于奖赏的脑机接口系统

基于奖赏的脑机接口系统,是用基于同一目标的奖赏来将大脑和计算机算法联系起来,也就是说,在整个三维空间中,给动物的奖赏(水或食物)和给算法的奖赏分布是同样的。这就给动物和计算机算法之间建立了对话通道,只要事先确定一个目标点,大脑和计算机都会根据奖赏来学习。

在强化学习中,算法根据环境的状态和回馈的奖赏选择一个动作来执行,环境根据执行的结果反馈奖赏。根据奖赏最大化的原则,即使事先不知道目标的位置,算法也能逐步学习到达目标(图 2A)。其中,空间中的奖赏分布事先通过 Q-learning 算法确定,Q-learning 算法的核心是评价函数,评价函数与环境状态和选择的动作有关。这里,环境不仅仅是指机械臂,还包括大脑,即大脑的活动决定环境状态,从而决定评价函数,算法根据评价函数选择一个最优动作,让机械臂执行,实现了大脑对机械臂的间接控制。

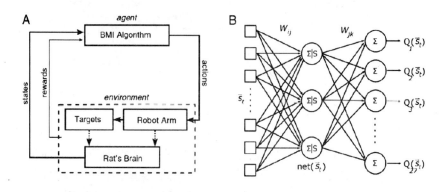

图 2 基于奖赏的脑机接口系统框架(A)和神经网络解码框架(B)

目前,我们选择神经网络做解码(图 2B),当然,任何一种算法都可以用在这里。我们构建了一个三层的神经网络,输入为 32 个通道的大脑活动状态,中间层包括五个中间节点,输出层为 27 个动作(动作集)各自的评价值。

系统(图 3)中有左、右两根压杆,各自有一个对应的 LED 灯,我们训练老鼠根据 LED 灯的提示,压相应的杆(左边的灯亮压左边的杆),任务完成就会在中间的奖赏中心反馈给老鼠一些水或食物。当老鼠学会后,就加上一个机械臂,机械臂根据上述神经网络的输出执行评价值最高的一个动作,我们希望老鼠根据 LED 的提示,操纵机械臂去压相应杆,任务完成,就会得到水或食物做奖赏。

实验中,我们发现,老鼠发现只靠"想"就能获得奖赏后,就会减少直至不

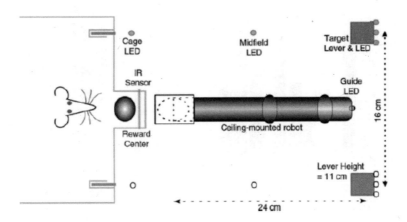

图3　系统框架示意图

再亲自压杆。老鼠学会任务后，我们在它脑部植入电极，分析实验结果，发现如图4所示结论。

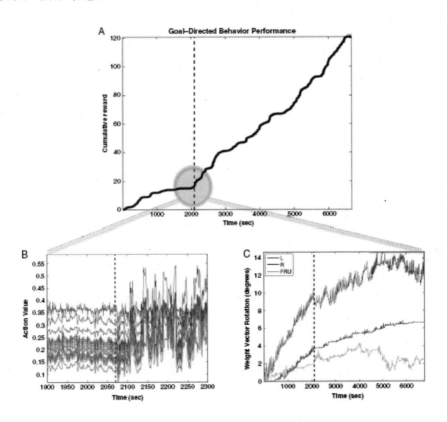

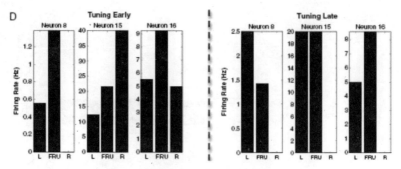

图4 学习速度和神经元表达的关系

(A) 试验中累积成功次数；(B) 动作的评价值随时间变化的函数；
(C) 三个最重要动作的权重变化；(D) 学习曲线拐点前后神经元调谐特性变化

图4显示的是第五天的实验结果分析。如图4A所示为实验成功的次数随着时间的增长曲线，主要反映出两个信息。在2000秒附近有一个突变点，曲线的斜率在突变点前后发生了巨大的变化，即突变点后，实验的成功率大大提高。首先这说明了系统的确能学会这项任务，其次我们之前已经发现神经元活动性编码信息的方式存在时变性，但我们通常假设这种时变性是一个缓慢的渐变过程，但图4A所示并不如此，它反映时变性过程存在突变点，在突变点附近神经元发放特性的变化是很快的。进一步观察神经元的发放特性变化，如图4D所示为三个比较重要的神经元在突变点前后编码三个动作的发放特性，可以看出存在明显的变化，说明动物在实验过程中的确在学习、适应系统。再来分析算法的变化：如图4B所示为可供机械臂选择执行的27个动作（如向左上移动、向右移动等），各自的评价值随时间变化的曲线（即神经网络算法的输出随时间变化的曲线）。可以发现一部分动作在突变点前后没有发生明显的变化，但是一部分动作在突变点前后发生了重大的变化。如图4C所示为神经网络算法中间层各节点权重（根据学习算法不断修正）与各自初始值（随机设定）的距离随时间变化的曲线（这里只选取了三个评价值最高的动作），可以看出在突变点前后，变化非常明显。两者共同说明了算法在实验过程中的确在学习、适应环境。因此，可以说这个系统中，大脑和算法都在共同学习，是共生性的脑机接口系统。

3 目标驱动的脑机接口系统

之前的脑机接口系统都由人事先指定目标的位置，如果能由动物自己来告诉我们它们想移动到什么位置就更好了。为了实现这一目标，我们不仅从主运动区获得信号,还从 NACC 区采集大脑活动,NACC 是一个位于沟回位置的核团,映射了奖赏中心的位置。利用 NACC 区的信号给出具体的评价函数,这也是一种强化学习的结构,我们称之为评价网络。评价网络的作用是通过解码 NACC 区的信号得到 NACC 区的状态,根据这个状态训练作用器,另外,作用器从主运动区获取信号,解码运动轨迹。这个系统成功的关键就在于 NACC 区信号是否能够真的反映奖赏中心的位置。系统框架如图 5 所示。

这里有许多数学公式,其中最重要的是评价函数,它的期望值依赖于奖赏给定状态和动作。一旦我们有了这个评价函数,我们就可以构造作用器的损失函数,所以关键问题是,我们如何从 NACC 区中获得评价函数。我们采用的方法是使用一个神经网络(图 6),当然任何非线性映射的算法都是合适的,而且现在我们确实也在使用着其他的一些结构,可以有效地避免局部最小值。一旦我们有了 NACC 区的信号,评价网络的输出是 -1 或者 1 作为奖赏,然后这个信号被用来训练动作器。

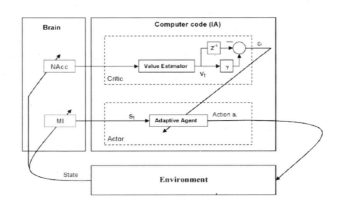

图 5　目标驱动的脑机接口系统框架

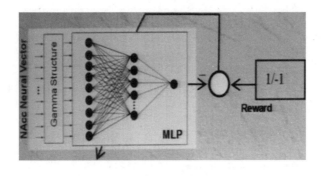

图 6　神经网络解码框架

　　在三只老鼠的实验中,我们大致发现了三种神经元,第一组神经元向两边移动时表现出兴奋性或抑制性,如图 7A、7B 所示;第二组神经元当向某一侧移动时有所反应,如图 7C、7D 所示;而第三组神经元对两个目标都没有响应,如图 7E、7F 所示。利用这些发放率的变化我们可以区分出神经网络反馈的奖赏(1 和−1),指导我们训练评价器,一旦评价器已经建立,就可以非常快速地训练执行器。

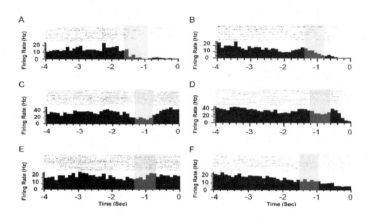

图 7　NACC 神经元发放特性

　　图 8 和图 9 有效地表明事先选择的神经信号成功地训练了评价器。因此可以看出,目标驱动的共生型 BMI 不需要一个训练集,因为它们总是在不停地用新的数据来进行学习。唯一必要的是设定哪个空间范围有奖赏。它们是互活应的,它们从实验者和环境中共同获得信息。因此我不认为老鼠是在模仿什么,它们是在试图通过神经调制来达到目标。

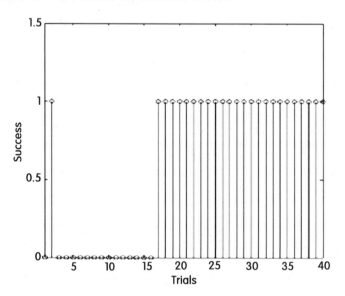

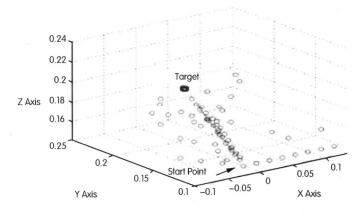

图 8　NACC 评价器指引下 M1 区神经元控制机械臂

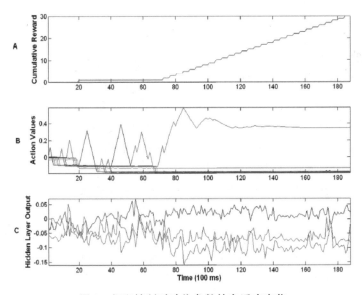

图 9　闭环控制时动作参数的自适应变化

4　总结

　　共生脑机接口系统减少了人为因素的限制,使得脑机接口更灵活,而不仅仅局限于训练集中的任务,可以从环境中学习,完成目标,是脑机接口研究未来所应当重视的一个发展方向。实现共生脑机接口系统有很多方法,其中有很多有趣的问题值得研究。

参考文献

［1］ Sanchez JC, Mahmoudi B, DiGiovanna J and Principe JC. Exploiting co-adaptation for the design of symbiotic neuroprosthetic assistants, *Neural Networks*, 2009, 22(3): 305 − 315

［2］ Mahmoudi B, Sanchez JC. A symbiotic brain-machine interface through value-based decision making, *PLoS ONE*, 2011, 6(3): e14760

讲座人简介

Jose C. Principe is Distinguished Professor of Electrical and Biomedical Engineering at the University of Florida, Gainesville, where he teaches advanced signal processing and machine learning. He is BellSouth Professor and Founding Director of the University of Florida Computational Neuro-Engineering Laboratory (CNEL). His research interests are centered in advanced signal processing and machine learning, Brain Machine Interfaces and the modeling and applications of cognitive systems. He has authored 5 books and more than 200 publications in refereed journals and book chapters, and over 380 conference papers. He has directed 65 Ph. D. dissertations and 67 Master's theses.

Dr. Principe is an IEEE and AIMBE Fellows a recipient of the INNS Gabor Award, the IEEE Engineering in Medicine and Biology Society Career Achievement Award, the IEEE Computational Intelligence Society Neural Network Pioneer Award, and Honorary doctor degrees from Universita Mediterranea, Italy, University of Maranhao Brasil, and Aalto University, Finland. He is Editor in Chief of the IEEE Reviews on Biomedical Engineering, Past Editor-in-Chief of the IEEE Transactions on Biomedical Engineering, current ADCOM member of the IEEE CIS society, IEEE Biometrics Council, and IEEE BME society, member of the Technical Committee on Machine Learning for Signal Processing of the IEEE Signal Processing Society; member of the Executive Committee of the International Neural Network Society, and Past President of the INNS. He is also a former member of the Scientific Board of the Food and Drug Administration, and a member of the Advisory Board of the McKnight Brain Institute at the University of Florida.

连接大脑的神经技术：挑战与机遇

Daryl R. Kipke

(Biomedical Engineering，University of Michigan，USA. NeuroNexus)

摘要：神经接口技术的进步为神经科学的研究提供了越来越强大的开发包，它涵盖了从实验设计、电极材料、电子元件到设备集成的一系列工具，极大地推动了神经科学的发展以及神经疾病的治疗。在微电极方面，除了硅基微电极的快速发展之外，新型的基于纳米结构材料的植入式微电极也正在开发之中，该电极可以获得高质量的慢性神经记录信号，同时能够减少对组织的损伤。另外，我们也正在开发一种多模态的神经探针，该探针在进行神经信号电记录的同时还能实现光刺激、神经化学信号记录和靶向药物释放等功能，可以更加全面地理解和认识神经回路。这些技术的进展无疑将使神经接口变得更为精确、稳定和可靠。

关键词：微电极阵列；神经探针；生物微机电系统；神经接口

Neurotechnology for Interfacing with the Brain: Technical Challenges and Emerging Opportunities

Daryl R. Kipke

(Biomedical Engineering，University of Michigan，USA. NeuroNexus)

Abstract：Advances in neural interface technologies are providing increasingly more powerful "toolkits" of designs, materials, components, and integrated devices to push the frontiers of neuroscience and treatments of neurological disorders. Beyond progressive improvements in silicon-based microelectrode arrays, new types of implantable microelectrodes are being developed that use advanced nanostructured materials to obtain high-quality chronic neural recordings from structures with a significantly reduced footprint compared to conventional implantable microelectrodes. Additionally, multi-modal neural probes are being developed to enable neural recording to be combined with optogenetics-based optical stimulation, neurochemical sensing, and/or targeted drug delivery in order to more fully engage neural circuits. These advanced technologies are extending the capabilities for precise, reliable, and high-fidelity neural interfacing in the brain.

Keywords: Microelectrode array, Neural probe, BioMEMS, Neural interface.

1　引言

据不完全统计,当今世界每年用在与大脑相关的神经疾病上的花费大约为2万亿美元,因此与大脑相关的神经技术是一个极其重要并且极具发展潜力的技术领域。目前,与大脑相关的各种神经疾病,有的已经能够实现临床治疗,有的正在被研究,如深部脑刺激是治疗帕金森病的主要手段之一。

神经工程是一个多种科技的融合交叉领域。神经科学、科学技术、临床需求三者相互影响,相互促进,共同推动着神经工程的发展。植入式神经探针作为一种与神经系统的接口,在探究生理过程中能提供非植入式方法无法达到的细胞级分辨率。当今的植入式神经探针能够同时记录组织内多个位点的神经信号。用于电极制造的衬底各式各样,包括硅、聚合物等。目前,为了使神经电极能够长期、稳定地记录到神经活动信号,大多数研究都集中在电极表面的特性研究上,如释放生物活性因子等。植入式神经探针作为与大脑的直接接触元件,负责采集大脑的神经信息,之后利用各种手段对采集到的信息进行处理,因此它的性能的好坏直接影响到整个系统的整体性能,它的发展对于神经科学的研究意义重大。

2　密歇根电极

微尺度神经电极是目前探究大脑机制的主要接口。临床应用的典型电极有大脑皮层电极和ECoG电极。ECoG电极的直径的典型值为5mm,深部脑刺激的电极的直径为1.2mm。随着当今科技的发展,微尺寸神经电极的尺寸越来越小,而且能够根据特定大脑的生理结构制造出合适的神经电极。光刻技术和硅腐蚀技术的发展为神经记录电极提供了一种新的制造工具,在硅衬底上使用平面光刻 CMOS 技术可以制造出能够同时记录多点的硅电极,该类电极相比于微丝电极,能够提供更多的采集位点,以及更高的空间分辨率。目前已经商品化、并被广泛应用的半导体集成硅微电极阵列主要有两大类,一类是美国犹他大学开发的针形微电极阵列,被称为 Utah 电极或平面电极;另一类是美国密歇根大学开发的线性微电极阵列,被称为密歇根(Michigan)电极或线性电极。本文主要介绍密歇根电极的种类、制造过程、性能测试及其商业化产品。

　　密歇根电极技术是一种制造二维平面电极的 MEMS 技术。该电极采用微电子制造技术,在硅为基底的薄片上,按照设计好的电极线路喷镀上导电金属;或者在整个覆盖有导电金属层的印制板上,蚀刻去除不需要的部分,留下需要的电极线路,导电金属可以是镍、不锈钢、钨、金或铂,然后,除了记录点以外,在其余连接记录点和输出端的导电线路上覆盖绝缘层,常用的绝缘材料是氮化硅,为了增强导电性能和生物相容性,记录点表面镀上铱或金。密歇根电极记录点的排列方式有很多种,最基本的排列结构是在一个记录杆上等间距线性排列一系列记录点,因此被称为线性电极阵列。只有单列记录点的电极也称为一维电极阵列,包含数根记录杆的电极称为二维电极阵列,数个二维电极组合使用还可以形成三维电极阵列。图 1 显示了一系列典型的密歇根电极。

**图 1　密歇根电极(左)在一个便士背面放了各种各样的硅电极(右)
四个 64 位点的探针组装成一个三维的结构**

　　密歇根电极的具体制作流程如图 2 所示[1]。首先以硅为衬底,通过深度硼扩散技术形成电极的针,然后覆盖一层绝缘层,之后覆盖一层导电层,再覆盖一层绝缘层,通过上升过程形成钛或钵的电极接触位点,最后通过腐蚀消融衬底,得到最终的神经电极。

　　除了上面提到的典型电极外,密歇根电极还能够根据特定的大脑组织结构制造出适于特定环境中应用的神经电极。如图 3 所示,该类电极的电极接触位点具有类似于四极管(tetrode)结构的排列方式,这种电极接触位点的紧

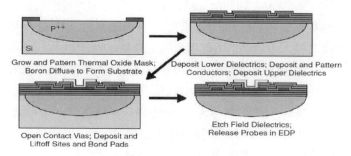

图 2　密歇根电极的具体制作流程

密排列方式能够在信号分析的时候有利于单个神经元的信号鉴别和分离。另外,有些电极不仅能够具有常见的信号记录和电刺激的功能,同时还具有传递物质的功能,如图 4 所示,这是一种存在传递通道的密歇根神经电极,该神经电极可以在刺激前实现靶向给药的功能。

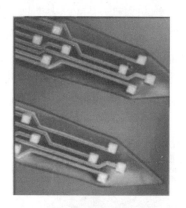

图 3　具有多位点的密歇根电极

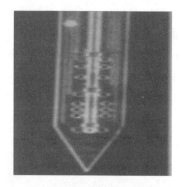

图 4　具有靶向给药功能的密歇根电极

　　理想的神经电极应该具有较小的体积,因而可以减少组织损伤。然而越多的电极接触位点能够记录到更多的神经元信息,体积的大小和电极接触位点的数量会相互制约。MEMS 技术的发展极大地改善了这一问题,因为在相同的组织损伤的情况下,该技术可以使得电极接触位点的数据急剧增加。目前,多点记录电极可以记录到多达上百个神经元的信息,图 5 示出了高密度密歇根阵列电极应用于大鼠的感觉皮层神经集群活动的记录[2]。该电极阵列由 8 根针组成,每根针上都具有 8 个电极接触点。图 5 - b 显示了该电极阵列各个记录位点记录到的局部场电位和神经锋电位信号。更为重要的是,电极记录点的精确定位也使得分析单个神经元之间的空间关系成为可能,这也是研究神经元集群的时空表示以及如何进行信息转换的前提。目前,增加电极接触位点的主要限制在于电极接触位点与颅外电子设

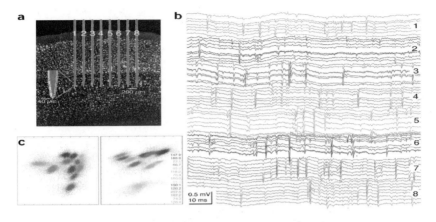

图5　老鼠感觉皮层神经元活动的高密度记录

(a) 在大脑的第五层放置具有 8 根探针的密歇根电极,每根探针的尖端边缘分布着 8 个电极接触位点,这些电极接触位点通过 $2\mu m$ 的连接线与颅外的电子设备相连接;(b) 一段原始信号记录,同时包括场电位和单神经元活动($1\sim5kHz$),在同一根探针上会有相同的锋电位,而不同的探针上的锋电位不同,这说明相隔大于 $200\mu m$ 的电极位点记录到的是不同的神经集群;(c) 来自同一根探针的神经元聚类,聚类点能够很好地分离说明能够更好地分离神经元

备连接的连接线宽度限制,如图 6 所示,线宽为 $2\mu m$,线间距也为 $2\mu m$,而工业生产可以提供 $0.18\mu m$ 线宽的制造工艺,希望在未来的几年里电极制造会采用更细的连接线,提供更多的电极接触位点,同时记录到更多的神经元集群信号。

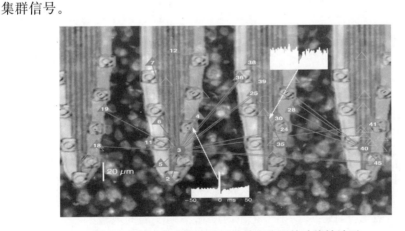

图6　老鼠感觉皮层记录到的神经集群的功能性地形

填充的三角形代表参与的锥体细胞,填充的圈代表中间神经元,未填充的三角形代表没有功能性连接的细胞,圈 35 与三角形 25 之间、圈 3 与三角形 2、6 之间、圈 40 与 45 之间的连线代表单突触抑制,圈 3 与三角形 4、12、25、36、45 之间的一条连线代表单突触兴奋,一条连线代表单突触抑制,其余连线代表单突触兴奋。从图中可以看出,大量的锥体细胞会激励中间神经元,而中间神经元会抑制局部和远端的一些锥体细胞。记录之间的垂直距离为 $20\mu m$,探针之间相距 $200\mu m$。白色的图形代表锥体细胞和中间神经元的相关图

依托于密歇根大学成立的 NeuroNexus 公司将密歇根电极技术转化为标准的商业化的神经接口产品,出售各式各样的密歇根电极产品,图 7 示出了该公司的各种产品。密歇根电极技术具有广泛的应用平台,针对不同的大脑部位,如浅部大脑、深部大脑以及皮层等,分别拥有特定的电极产品。另外,光基因技术作为神经工程领域中的新兴的研究热点而备受关注,研究光基因技术当然离不开与光有关的一套实验设备,NeuroNexus 公司研制出能够进行光传递的神经电极,该电极是目前最早商业化的光传递神经电极。

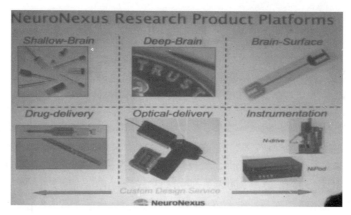

图 7　NeuroNexus 公司的各种产品

3　神经电极特性研究

目前,微尺度密歇根神经探针的研究重点主要在于两个方面,一方面是神经探针的多模态研究,能够提供电记录和电刺激的功能,或者能够提供化学感知以及药物传递的功能,另一方面是微尺度神经探针的长期稳定性研究,能够长期稳定地记录到高质量的神经信号,同时能够长期准确地进行刺激。

为了研究神经探针的长期有效性,我们将神经微电极植入到豚鼠大脑中,观察神经电极上的有效位点的数目变化[3]。图 8A 中给出了能够记录 9 周以上的电极的有效情况,图 8B 给出了记录时间达不到 9 周的电极的有效情况。以图 8 中可以看出,记录时间能够达到 9 周以上的电极,能够保持记录的有效位点数目变化较为稳定,且维持在 80% 左右,而那些记录时间达不到 9 周的电极,它们的有效位点数目会随时间的推移逐渐变差,甚至完全失效。为何在动物实验中,有些电极能够维持较为稳定的信号记录,而另外一些电极不能维持稳定的信号记录呢? 具体原因尚且不得而知,还有待于进一步研究。

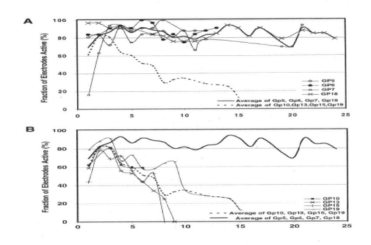

图8　有效电极位点的数量作为植入后周期数的函数

（A）长期记录(大于9周)。两图中的实线是记录时间大于9周的几只老鼠的有效电极位
点的平均。（B)短期记录(小于9周)。两图中的虚线是记录时间小于9周的几只老鼠的
有效电极位点的平均

　　神经电极的植入必然会引起神经组织的损伤,同时会引起组织的生物相
容性反应[4]。图9显示了电极植入点周围组织的生物反应。随着时间的推
移,植入点周围的神经胶质细胞越来越多,最终会将电极包裹起来,将电极与活
跃的神经元相分离。神经胶质细胞增生引发一系列的问题,如大脑如何对植入
式神经电极做出反应? 相关组织反应的原因和机制是什么? 组织反应和组织
功能的关系是怎样的? 多大程度的组织反应是可以忍受的? 电极如何改进可
以提高电极的有效期,同时提高信号的质量? 这些问题都有待进一步研究。

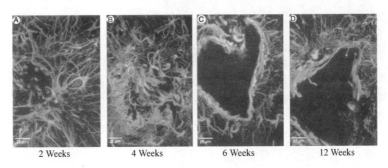

图9　通过 GFAP 着色显示的角质疤痕的形成过程

　　在第2周和第4周时,星形角质细胞重新填充电极移除后留下的空隙,在第4周时,在植入点
500μm 处,星形角质细胞大量增生,角质疤痕在第6周后基本形成,厚度约为 $50\sim100\mu$m

Ω有人提出了一些假设,试图解释在电极植入后,大脑组织是如何进行反应的。为了研究神经组织对植入式神经电极的组织反应,Patrick A. Tresco 教授[5]将电极植入到老鼠的大脑皮层中,使用定量方法来分析2周、4周以及12周后组织对植入电极的反应。通过使用 ED-1 对小胶质细胞/巨噬细胞进行着色,分析 ED-1 的密度即对小胶质细胞/巨噬细胞进行分析。图 10A 显示了 ED-1 免疫反应区是圆形的,这反映出植入电极的几何形状,沿着整个电极轴是对称的。图 10B 说明在离电极较远的大脑区域没有明显的小胶质细胞/巨噬细胞增生,平均的 ED-1 的密度作为离电极组织接口的距离的函数,观察2周、4周以及12周后的 ED-1 密度情况(图 10C～图 10E),在2周的时候,ED-1 的免疫反应在距离接口处较近的区域比较剧烈,等到4周和12周后,ED-1 的染色还是集中于靠近电极组织接口处,定量分析曲线下的区域面积,在 $100\mu m$ 处几乎没有统计上的差别,不过,ED-1 的平均荧光信号随着时间的推移,变化越来越剧烈。

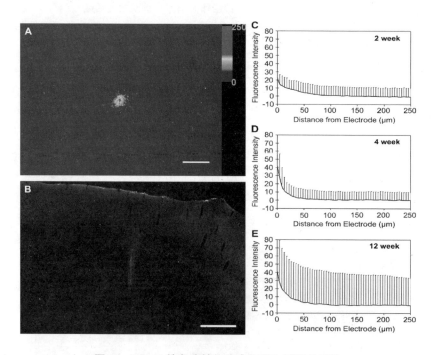

图 10 ED-1 的免疫性作为电极植入时间的函数

使用 GFAP 对星形角质细胞进行着色,图 11A 显示了包裹这 ED-1 免疫区的 GFAP 免疫区,与 ED-1 相似,GFAP 免疫区也是圆形的,反映出植入电极的几何形状,横切面为圆形,冠状切面为圆柱形(图 11B),图 11C～图 11E 显示了在2周、4周以及12周后的 GFAP 免疫性的定量分析,从图中可以看出,2周时,紧靠电极组织接口的免疫性比无植入电极时会减弱,4 周和 12 周后,免疫性会有所增加。在距离电极 $100\mu m$ 处的免疫性最强,即星形角质细胞的数量最多,与 ED-1 类似,GFAP 的免疫变化性随着时间的推移而增加。

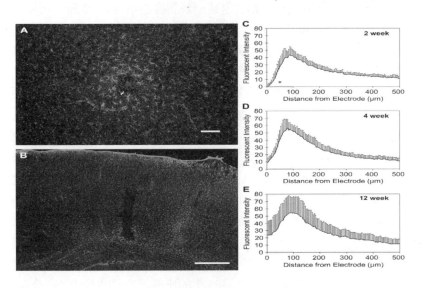

图 11　GFAP 免疫性作为电极植入时间的函数

使用 NeuN 对神经细胞进行着色,图 12A 出示了植入电极 4 周后的 NeuN 成像情况,图 12C~图 12E 画出了 2 周、4 周以及 12 周的定量分析相对于无损伤的组织,在距离电极组织接口 $50\mu m$ 范围内的 NeuN 密度有明显的

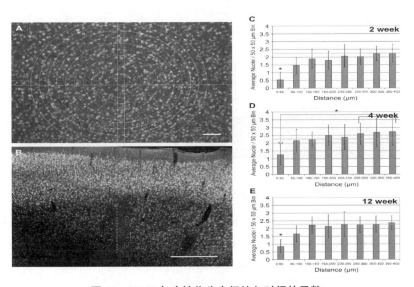

图 12　NeuN 免疫性作为电极植入时间的函数

减少,图 12B 显示了在距离电极组织接口较远的组织区域,神经元的数量基本没有明显的变化。与前两种细胞的情况相似,神经元的数量变化随着时间的推移而增加。

为了使得神经电极能够长期稳定地记录到神经信号,不少研究者使用导电聚合物涂层来修改电极接触位点,以此提高信噪比,增加电极的生物相容性。Ludwig 等人研究了将电极接触点利用乙烯基进行材料修饰[6],并对经过修饰的电极与未经修改的电极的性能进行了比较。首先,对比研究电极记录位点的阻抗性能。由于动作电位的频带大多集中于 1kHz,因此通过测量电极记录位点在 1kHz 的阻抗来评价记录探针的性能。从图 13 中可以看出电极记录位点在 1kHz 时的阻抗分为三个不同的时间阶段,图中短线代表 64 个电极记录位点的阻抗的标准差,PEDOT 代表修改后的电极,CONTROL 代表未修改的电极。在手术后的前三天里,电极记录位点的平均阻抗保持相对稳定。未修改的电极位点的平均阻抗为 $0.98M\Omega\pm0.08M\Omega$,而修改后的电极位点的平均阻抗为 $0.13M\Omega\pm0.06M\Omega$,大约 7：1 的比例,修改后的电极位点的阻抗变化也相对于未修改的电极位点要小。从第 3 天到第 15 天可看成第二个阶段,电极阻抗有明显的增加,未修改的电极位点的平均阻抗为 $1.7M\Omega\pm0.9M\Omega$,而修改后的电极位点的平均阻抗为 $1.1M\Omega\pm0.8M\Omega$,大约 1.5：1 的比例。15 天之后为第三个阶段,电极位点的阻抗逐渐下降。未修改的电极位点的平均阻抗为 $1.5M\Omega\pm0.3M\Omega$,而修改后的电极位点的平均阻抗为 $0.81M\Omega\pm0.3M\Omega$,大约 2：1 的比例。电极阻抗的变化趋势也与组织的免疫反应密切相关。

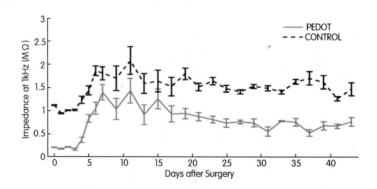

图 13　在 1kHz 时电极记录位点随时间推移的平均阻抗

Ludwig 等人还对比研究了电极记录的信噪比性能。从图 14 中可以看出,随着时间的推移,记录到的神经信号的信噪比可以分成三个时间阶段。在手术后的两天里,修改了的电极记录到的神经信号的平均信噪比为 5.1±1.2,而未修改的电极记录到的神经信号的平均信噪比为 4.1±1.1。在手术后第 3 天到第 15 天的这段时间里,两者记录到的信号的信噪比大致相同,修改了的电极记录到的信噪比为 4.8±1.8,未修改的电极记录到的信噪比为 4.9±1.7。15 天后,两者的信噪比再次出现明显的差异,修改了的电极记录的信噪比为 5.1±1.2,未修改的电极记录到的信噪比为 4.3±1.0。总体看来,修改后的电极记录到的神经信号的信噪比要好于未修改的电极。

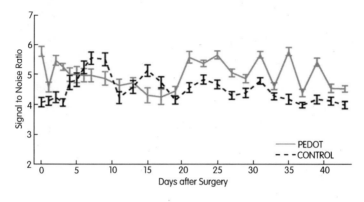

图 14　随着时间的推移电极记录信号的平均信噪比

4　总结

　　植入式神经电极是探究大脑神经系统生理过程的一种重要工具,通过分析它所记录到的神经信号,来解释大脑的生理过程。因而长期稳定的神经信号记录,对于神经科学的发展尤为重要。针对如何才能记录到长期有效的高质量信号,我们总结出以下五点:① 尽量减少组织损伤;② 电极具有高质量的电学特性;③ 电极具有更小的体积;④ 电极的柔韧性更好,能够更加贴切地靠近大脑组织;⑤ 电极的表面材料具有生物活性。随着科学的发展和制造工艺的进一步提高,神经电极的生物相容性问题一定会得到有效的解决。植入式针型神经微电极技术作为神经-电子接口,必将为我们探索神经系统和大脑的奥秘以及寻求治疗神经疾患的有效手段提供更为有力和灵活的工具。

参考文献

[1] Wise KD, Anderson DJ, Hetke JF, *et al*. Wireless implantable microsystems: High-density electronic interfaces to the nervous system. *Proceedings of the Ieee*, 2004, 92(1): 76-97

[2] Buzsaki G. Large-scale recording of neuronal ensembles. *Nature*

Neuroscience, 2004, 7(5): 446 – 451

[3] Williams JC, Rennaker RL, and Kipke DR. Long-term neural recording characteristics of wire microelectrode arrays implanted in cerebral cortex. *Brain Research Protocols*, 1999, 4(3): 303 – 313

[4] Cheung KC. Implantable microscale neural interfaces. *Biomedical Microdevices*, 2007, 9(6): 923 – 938

[5] Winslow BD, and Tresco PA. Quantitative analysis of the tissue response to chronically implanted microwire electrodes in rat cortex. *Biomaterials*, 31(7): 1558 – 1567

[6] Ludwig KA, Uram JD, Yang JY, *et al*. Chronic neural recordings using silicon microelectrode arrays electrochemically deposited with a poly(3,4 – ethylenedioxythiophene) (PEDOT) film. *Journal of Neural Engineering*, 2006, 3(1): 59 – 70

讲座人简介

Daryl R. Kipke, PhD. is a professor in the Department of Biomedical Engineering in the College of Engineering at the University of Michigan (Ann Arbor, Michigan USA). Dr. Kipke leads the Neural Engineering Lab at Michigan and also directs the Center for Neural Communication Technology, a NIH-supported Biotechnology Research Center. Dr. Kipke has an internationally recognized research program in neural engineering, neuroprostheses, neural interfaces, and neural biomaterials. Dr. Kipke is also the co-founder and President/CEO of NeuroNexus Technologies, Inc. (Ann Arbor, Michigan; http://NeuroNexus. com), a neurotechnology company providing advanced brain interface devices for neurological and scientific applications. Previously, Dr. Kipke co-founded and helped to lead Neural Intervention Technologies, Inc. from a University of Michigan startup in 2001 to acquisition by W. L. Gore in 2006. Dr. Kipke is a Fellow of the American Institute of Medical and Biological Engineering and a member of several neuroscience and biomedical engineering societies.

光学脑功能成像研究和成果转化

Banu Onaral

(School of Biomedical Engineering, Science and Health Systems,
Drexel University, Philadelphia, 19104 USA)

摘要：基于近红外光谱技术（NIRS）的光学成像系统是一种广泛应用于脑功能研究的非植入式方法。NIRS 通过检测氧合血红蛋白和脱氧血红蛋白的浓度来间接监测大脑的活动。Drexel 大学的脑光学成像研究团队开发了基于 NIRS 的脑功能监测系统，用以评估健康人和病人的认知活动。该系统具有便携式、安全、价格低和无创等优点，能在多种场合下研究大脑皮层的活动情况，可实现多方面的应用，包括工作绩效评价、麻醉状态监测、神经康复、脑机接口、心理健康治疗等。Drexel 大学的脑光学成像研究团队鼓励跨地区跨国界合作，这个组织全称是 Cognitive Neuroengineering and Quantitative Experimental Research (CONQUER) Collaborative，致力于功能成像技术的产业化。

关键词：近红外光谱；脑功能成像；认知神经科学；产业化

Translational Research in Functional Optical Brain Imaging

Banu Onaral

(School of Biomedical Engineering, Science and Health Systems,
Drexel University, Philadelphia, 19104 USA)

Abstract：Near-infrared spectroscopy (NIRS) based optical imaging systems have been widely used in functional brain studies as a noninvasive tool to study changes in the concentration of oxygenated hemoglobin (oxy-

Hb) and deoxygenated hemoglobin (deoxy-Hb). Based on the NIRS technique, Drexel University's Optical Brain Imaging team has developed a functional brain monitoring system (fNIR) to assess cognitive activity of healthy subjects and patients. The fNIR is a portable, safe, affordable and negligibly intrusive monitoring system which enables the study of cortical activation-related hemodynamic changes under various field conditions. This presentation will provide an overview of applications of the fNIR including human performance assessment, depth of anesthesia monitoring, neuro-rehabilitation, brain computer interface for locked-in patients, mental health applications with special focus on "brain-in-the loop" applications in motor learning and robotic rehabilitation. The audience will be introduced to the *Cognitive Neuroengineering and Quantitative Experimental Research (CONQUER) Collaborative* which hosts the Optical Brain Imaging team and welcomes all regional, national and international partnersdedicated to the research, development, integration, translation, productization and commercialization of functional imaging techniques to monitor human brain activation.

Keywords: NIRS, Brain functional imaging, Cognitive neuroengineering, Industrialization

1 引言

近年来,随着科学技术的迅速发展,认知神经科学领域涌现了一批功能强大的脑成像技术,如功能性磁共振技术(fMRI)、正电子发射层析技术(PET)、单光子发射计算层析技术(SPECT)、自发脑电活动(EEG)、事件相关电位(ERPs)和脑磁图(MEG)等。这些技术大大拓展了认知神经科学领域的研究。但是上述设备大多局限于实验室环境,难以在包含大肌肉群的运动项目中使用,难以考察各种条件下(如运动条件)的脑神经机制。而近红外光谱技术(NIRS)则可弥补这一不足,近红外光谱技术以氧合血红蛋白、脱氧血红蛋白和细胞色素氧化酶等为指标,考察与神经元活动、细胞能量代谢以及血液动力学相关的大脑功能,所以也称近红外功能成像技术(fNIR)。这一技术具有时空分辨率较高(图 1)、便携性强、价格低廉和无创等优点,在认知神经科学和医学等研究中得到了越来越广泛的应用。

2　fNIR 技术

fNIR 是一种间接检测大脑活动的技术,这种技术手段与 fMRI 有些相似,都是检测血红蛋白的变化。神经元需要能量来产生电活动,这些能量来自葡萄糖和氧气等分子的代谢反应,因此一群神经元的激活需要相当量的氧气。

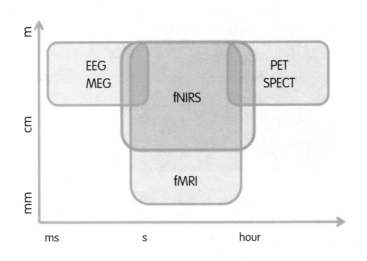

图 1　几种脑功能成像技术的时空分辨率比较

这些氧气原来依附于氧合血红蛋白。这种蛋白将氧气从血液中传输到神经细胞内部后,就转化成脱氧血红蛋白。所以氧合血红蛋白和脱氧血红蛋白的含量能够表征脑部的活动。这些含量的变化可以通过近红外光在颅内的吸收和散射来检测,具体计算可利用修正后的 Beer-Lambert 定律。假设 G 是测量因子常数,I_0 为入射光强度,I 为出射光强度,α_{HB} 和 α_{HBO_2} 分别为脱氧血红蛋白和氧合血红蛋白的摩尔吸光系数,C_{HB} 和 C_{HBO_2} 分别为脱氧血红蛋白和氧合血红蛋白的浓度,L 为光通过的路径,是吸收系数和散射系数的函数,我们可得:

$$I = GI_0 \, \mathrm{e}^{-(\alpha_{HB}C_{HB} + \alpha_{HBO_2}C_{HBO_2})L}$$

一般来说,fNIR 设备由两部分组成,即光源和检测器。光源发射近红外光进入大脑,经大脑组织散射返回后由检测器检测(图 2)。根据光源的输入特征,fNIR 主要有三种类型:

（1）连续波动：入射光源的强度在测量过程中保持恒定，通过出射光的强度变化来检测脑部活动；

（2）频率调制：入射光的强度由一个正弦波调制，通过波形的强度和相位变化检测脑部活动；

（3）短时脉冲：入射光是一个个可达皮秒级的脉冲，通过脉冲响应检测脑部活动。

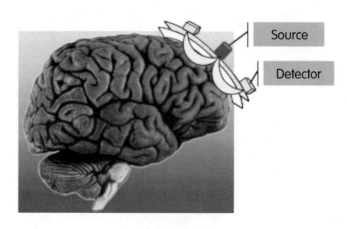

图 2　fNIR 设备原理图

fNIR 检测的范围不如 fMRI 深广，检测部位由设备安放部位确定。但 fNIR 的便携式特点使其可以测量人在日常环境中的脑部活动，这是 fMRI 做不到的。目前已有很多 fNIR 的产品，比如日立 ETC-4000 CW 系统、ISS 公司的 IMAGENT 系统、UCL 的 32 通道 MONSTIR 系统。

3　应用

Drexel 大学在 2001 年 8 月开发出了基于连续波动的便携式 fNIR 设备，之后不断对其加以改进。目前的设备可以检测整个头部，可以无线传输，并已经和公司合作生产设备，主要应用包括以下几个方面：

3.1　人的工作绩效研究

典型例子是便携式脑部成像在飞行控制上的应用，可以用于监控航空路线控制人员的脑部活动和压力承受情况、飞行员在虚拟飞行中的学习情况和地面操作人员的训练和压力承受情况，保证飞行安全。这种技术被用于美国航空管理局新一代的航空交通系统中，比如监测调度人员不同工作强度下的大脑活动(飞机数目＝6,12,18)。fNIR 可以用于空军的无人驾驶飞机研究。对于无人驾驶飞机,地面操作人员的技术水平,认知相关工作的强度,对各种情况的敏感程度都是关键。60%～80%的驾驶失败是由人为错误造成的,而这些错误主要是因为工作强度的增加和训练不充分。用 fNIR 可以客观监测受训者的学习进度,包括从新手到熟练工的过程,以及实时监测大脑的工作强度。

3.2　大脑整合学习

Drexel 大学与以色列 Edmond J. Safra Brain Research Center 合作展开语言学习和学习障碍的研究,包括研究语言学习中的大脑活动、区分阅读正常者和阅读障碍者的大脑活动和改善学习能力的研究。

3.3　临床应用

为检测脑外伤的血肿情况而开发的仪器已被全世界范围内的医院采用;监测脑外伤中认知受损和恢复情况;与医生合作将 fNIR 应用于检测麻醉的程度,尽量减少麻醉剂量,同时又可防止病人从手术中苏醒,目前处于临床前(pre-clinical)阶段;老年人的认知监测,这种监测可以在日常生活中进行,比如行走;研究如何将 fNIR 用于自闭症和精神分裂症等大脑疾病的诊断;对于瘫痪病人,开发脑机接口技术;新生儿脑部活动的无线监控。

4　总　结

借助近红外光谱技术,考察大脑皮层的血流动力学特征,有助于揭示大脑皮层的功能性激活与认知的关系及其机制。目前 fNIR 已有多方面的应用和商业化产品。而 Drexel 大学 fNIR 的开发得益于该大学名为 CONQUER CollabOrative 的组织,该组织强调多学科交叉、国际合作以及技术产业化。提出开发一个医疗仪器,不仅要做产品开发,还要做商业价值评估,要多方合作才能成功,而大致过程可分为基础和应用研究、原型开发、临床应用原型开发、临床预实验、临床验证、工业产品原型、产业化等步骤。这个过程和模式已经在 fNIR 中取得成功,可以预见将在高校研究成果的产业化过程中发挥更大的作用。

参考文献

[1] Heeger DJ, Ress D. What does fMRI tell us about neuronal activi-

ty? *Nature Reviews Neuroscience*, 2002, 3(2): 142 - 151

[2] Obrig H, Wenzel R, Kohl M, Horst S, Wobst P, Steinbrink J, Thomas F, Villringer A. Near-infrared spectroscopy: does it function in functional activation studies of the adult brain? *International Journal of Psychophysiology*, 2000, 35(2 - 3): 125 - 142

[3] Izzetoglu M, Bunce SC, Izzetoglu K, Onaral B, Pourrezaei K. Functional brain imaging using near-infrared technology. *IEEE Engineering in Medicine and Biology Magazine*, 2007, 26(4): 38 - 46

讲座人简介

Dr. Banu Onaral is a H. H. Sun Professor of Biomedical Engineering and Electrical Engineering at Drexel University, Philadelphia, PA. Dr. Onaral joined the faculty of the Department of Electrical and Computer Engineering and the Biomedical Engineering and Science Institute in 1981. Since 1997, she has served as the founding Director of the School of Biomedical Engineering Science and Health Systems.

Her academic focus both in research and teaching is centered on information engineering with special emphasis on complex *systems and biomedical signal processing in ultrasound and optics*. She has led major research and development projects sponsored by the National Science Foundation (NSF), National Institutes of Health (NIH), Office of Naval Research (ONR), DARPA and Department of Homeland Security (DHS). She supervised a large number of graduate students to degree completion and has an extensive publication record in biomedical signals and systems. She is the recipient of a number of *faculty excellence awards* including the 1990 Lindback Distinguished Teaching Award of Drexel University, the EDUCOM Best educational Software award and the NSF Faculty Achievement Award.

Dr. Onaral's translational research efforts for rapid commercialization of biomedical technologies developed at Drexel and its partner institutions have resulted in the creation of the Translational Research in Biomedical Technologies program. This initiative brings together academic technology developers with entrepreneurs, regional economic development agencies, local legal, business and investment communities. Under her leadership, the program has been recognized by the *Coulter Translational Research Partnership award*. She is currently leading the creation of the regional Translational Research Partnership Institute to pool regional resources and assets to commercialize health care solutions.

Dr. Onaral's professional services include chair and membership on advisory boards and strategic planning bodies of several universities and funding agencies, including service on the National Science Foundation's Engineering Advisory Board, and on the proposal review panels and study sections. She participated in the strategic planning team charged with the creation of Sabanci University established in 1998 in Istanbul, Turkey and served on its board of trustees.

Her professional responsibilities have included service on the Editorial Board of journals and the CRC Biomedical Engineering Handbook as Section Editor for Biomedical Signal Analysis. She served as *President of the IEEE Engineering in Medicine and Biology Society* (EMBS), *the largest member-based biomedical engineering society in the world*. She organized and chaired the 1990 Annual International Conference of the EMBS and Co-Chaired the 2004 Annual Conference of the Biomedical Engineering. She is a *Fellow of the IEEE* Engineering in Medicine and Biology Society, the American Association for the Advancement of Science (AAAS) and *a Founding Fellow of American Institute for Medical and Biological Engineering* (*AIMBE*). She served on the inaugural Board of the AIMBE as publications chair and as Chair of the Academic Council and as the President of the Turkish American Scientists and Scholars Association.

稳态视觉诱发电位在脑机接口中的应用

高上凯

(神经工程实验室,生物医学工程专业,医学院,清华大学,北京,中国)

摘要:稳态视觉诱发电位(SSVEP)是大脑对周期性视觉闪烁刺激产生的反应。SSVEP 已经成功应用于脑机接口(BCI)和认知任务研究中。这次讨论将介绍各种基于 SSVEP 的脑机接口(BCI),并强调其中的时域、频域、相位和空间域的信号分析方法,同时介绍 SSVEP 在认知上的应用。

关键词:稳态视觉诱发电位;脑机接口;认知

The Applications of Steady State Visual Evoked Potentials in Brain-computer Interfaces

Shangkai Gao

(Lab of Neural Engineering, Dept. of Biomedical Engineering, School of Medicine, Tsinghua University, Beijing, China)

Abstract: Steady state visual evoked potential (SSVEP) are brain responses induced by periodic flickering visual stimuli. SSVEPs have been successfully applied in brain-computer interfaces (BCI) and cognitive study. In this talk, the various designs of SSVEP-based BCI are introduced, in which the signal analysis methods in frequency, phase, time, and space domains are emphasized. Meanwhile, some of the SSVEP applications in cognitive study are also presented.

Keywords: SSVEP, BCI, Cognition

1 引言

当外界视觉刺激以高于 6Hz 的频率闪烁时,在大脑皮层视觉区会诱发出明显的电位变化,这就是由视觉刺激诱发的与外界刺激频率相同的 SSVEP。

现在 SSVEP 已经成为应用于脑机接口中非常有效的交流媒介。在 SSVEP-BCI 系统中有一些同时在闪烁的目标,但是各个目标的闪烁频率不同,通过计算得到的 SSVEP 信号频率,就可知被试注视的是哪个目标,从而可以转变为外部设备的控制命令。

此外,通过查询 1999 年至 2010 年 EI 收录的 SSVEP-BCI 相关文章(查询关键字是 BCI 和 SSVEP),我们惊讶地发现,这类文章的数目快速增长,尤其是近三年,几乎成指数增长,这说明基于 SSVEP 的 BCI 已经越来越得到重视。因为现有的 SSVEP-BCI 应用已有很多成果,所以本次报告主要介绍 SSVEP的应用,而后对本实验室研究组做的一些工作进行简要介绍,最后介绍一下 SSVEP 在其他方面的应用和未来需要解决的问题。

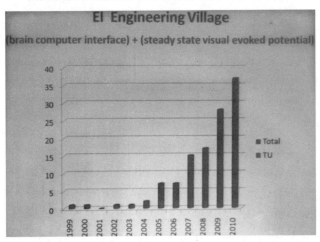

图1　1999—2010 年 SSVEP-BCI 文章增长图

2 SSVEP 应用于 BCI

现今,很多研究已经成功将 SSVEP 应用于各种 BCI 系统,主要应用包括机器人控制、轮椅控制、电动假肢控制、光标控制、拼写任务、谷歌搜索、玩游戏等等。此外,基于 SSVEP 的低功耗 BCI,混合 BCI 也已有成果。下面,就其中的一些应用做简要介绍。第一个应用例子是机械手控制。日本东京大学的研究改变了传统的闪烁棋盘的实验范式,改用闪烁的情感视频,从而达到增强 SSVEP 的响应。被试通过 SSVEP 选取电脑屏幕上的目标,设备就可以转化成各种控制命令,控制机器手做出被试想要的各种动作[1]。

第二个例子是不莱梅大学的研究,用 SSVEP 控制半自动康复机器人系统。被试坐在电动轮椅上,这个系统利用被试的 SSVEP 信号使四肢瘫痪康复机器人控制高层次的命令系统(如"抓握瓶子"),然后介入执行要做的任务,实现了辅助一些瘫痪病人的作用[2]。

第三个例子是 Graz 大学研究组,首先利用虚拟现实技术搭建一个阿凡达导航系统,使被试在不同的虚拟场景中通过 SSVEP 选择目标方向,从而在整个虚拟房间中漫游[3]。

第四个例子是圣埃斯皮里图州联邦大学的研究成果,被试面对的屏幕界面上有四个条纹(上、下、左、右四个方向)的闪烁刺激,被试通过选择这四个条纹,激发 SSVEP 信号,由系统转换为控制命令自主地控制轮椅四个方向的运动[4]。

此外,利用 SSVEP 信号控制电动假肢,在假手的下边安有四个 LED,从而控制假手的向左转、向右转、打开和握紧;SSVEP-BCI 用于控制鼠标移动;SSVEP-BCI 用于谷歌搜索(图 2A);用于拼写;与功能性电刺激相结合;清华大学和上海交通大学共同完成的 SSVEP-BCI 轮椅控制(图 2B)等等。此外,还有基于 SSVEP 的混合 BCI 以及低功耗 BCI。总之,所有这些应用都是 SSVEP-BCI 在实际应用的成功示例。但如果只看到这些就觉得 SSVEP 简单那就错了。迄今,SSVEP 的产生机理还是未知的,SSVEP 和正在发展的 EEG 之间的关系也不清楚。所以,无论在生理记忆科学还是应用科学,都还有很多问题值得我们更深入的研究。

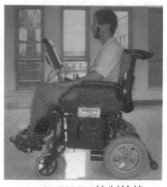

A. SSVEP用于谷歌搜索　　　　　　B. SSVEP控制轮椅

图 2　SSVEP-BCI 应用

3 系统设计及本实验组的工作

上面主要介绍 SSVEP 在脑机接口中的应用,但是这个系统是怎样实现的,以及实现过程中需要注意的问题是什么? 下面就本实验组所做的研究工作做一下介绍。

3.1 SSVEP 信号的特征

SSVEP 信号的特征包括频域、相位、空间、时域四个方面(图 3)。给予一定频率的视觉刺激,视觉皮层会产生相应的响应,这就是 SSVEP,它的频率和刺激信号的频率相同。

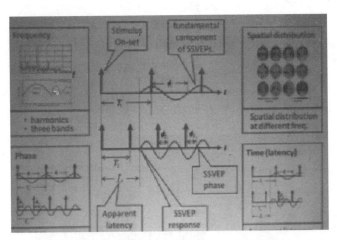

图 3 SSVEP 信号的频域、相位、空间、时域特征

3.1.1 频域的特征

在这里介绍两个重要的特点:

(1) 和声学的基本原理,这里有一点需要注意:当刺激以一定频率闪烁,产生的 SSVEP 信号不仅包含基频成分,还会有谐波成分。而且,有时二次谐波成分要比基频成分的幅值还要高。即使刺激是单一频率的也可能会导致谐波的产生。

(2) SSVEP 的最重要的频段:大部分人有三个频段:低频,中频,高频。它们对刺激都很敏感。

3.1.2 相位特征

由于 SSVEP 是有意识的响应,所以 SSVEP 是与刺激信号锁相的,这对于 SSVEP 的设计也是很重要的。

3.1.3　空间特征

信号在视觉皮层的是有空间分布的，这对于 SSVEP-BCI 系统的设计也是很重要的。

总之，频域、相位、空间、时域的特征对于 SSVEP 系统的设计都是十分重要的。我们应该综合考虑和运用各种特征，最终快速准确地识别目标。

3.2　实验范式

下面介绍本实验组的实验范式，主要有 3 种实验范式。

3.2.1　幅频谱实验范式

这种实验范式主要是利用信号的频域特征。当被试注视以一定频率闪烁的目标时，通过计算信号频谱，就能在频谱图中看到刺激频率段和谐波段有明显的峰值。通过这个特点，可以确定被试注视的目标。

此系统的特点（见图 4）包括：①非植入式技术；②很少数目的电极，只有三个通道，一个接地，一个作为参考，另一个是信号；③较高精度，较少训练，方便易用；④可以用较多数目的目标（图 5 实验范式就是用了 12 个目标），⑤用户舒适（较弱的光线）；⑥系统简单而且便宜；⑦使用方便，较少训练就能得到较高精度；⑧高信息传输率（约 50bit/min）。所以这种范式应用比较广泛。

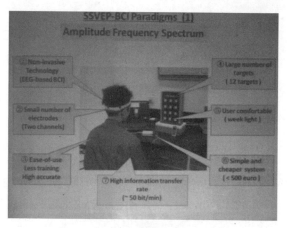

图 4　SSVEP 的实验范式及其优点

这个 SSVEP-BCI 系统看起来很简单，但是也需要做很多的工作，比如频带选择、通道选择、谐波权衡等等。

3.2.2　幅度＋相位的频谱范式

这也是一个通用的 SSVEP-BCI 范式，可用于这个范式的频段通常是有限的频段，在本研究组我们常用的频段是 10～20Hz。

　　这个范式综合运用频谱特征和相位特征。它可以解决这样的问题：对于有限数目的刺激频率，例如只用一种刺激频率时，我们怎样区分较多的目标？

　　解决方案就是用锁相特征（脑电信号和刺激信号的锁相值），也就是说我们可以用同一频率但带有不同相位的刺激信号作为不同的目标刺激，这样我们就利用频谱特征和相位特征共同区分不同的目标。

3.2.3　注意力范式的 SSVEP-BCI

　　有两种注意力的定义，空间注意力和非空间注意力。空间注意力就是被试头部固定屏幕中央，然后，被试再将注意力集中在这边或者那边不同的目标处。非空间注意力就不用固定被试头部，只需要直接正视正前方来注意屏幕上的目标即可。假设是非空间注意力，如图 5，这里有灰白两种颜色的点，当被试注视灰色的点时，会产生 10Hz 的响应峰值，当被试注视白色的点时，将会产生 12Hz 的响应峰值。

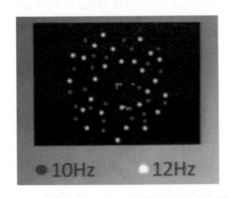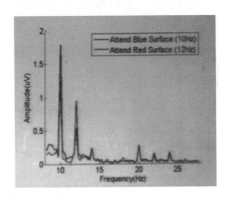

图 5　注意力范式的 SSVEP 响应

4　SSVEP 在认知神经科学中的应用

　　"双眼竞争"是一个众所周知的经典的用来研究意识视觉感知的实验范式。例如，两幅不同的图片出现在被试面前，被试左眼只能接收左边的图片，右眼只能接收右边的图片，但当两幅图片同时出现在被试眼前时，被试一次只能接收一张图片。至于接收哪张图片，只有被试自己才能选择决定，外界是无法区分的。所以可以采用一种方法来进行被试主观选择的验证，例如左边的图片闪烁频率是 30Hz，右边图片以不同的频率闪烁。被试主观地选择左边的图片并通过按钮来表示他选择的是哪个图片，同时记录脑电信号，并看视觉皮层哪个频率成分得到了加强。通过验证得出，如果选择的是左边的图片，那么视觉皮层脑电信号中 30Hz 的信号得到加强，从而实现了验证。

5　总结

　　此次报告结合本实验组的研究，首先将现有的 SSVEP-BCI 应用成果进行了一些介绍，并提出在系统设计以及信号分析过程中需要注意的问题，指出信号频域、相位、时间潜伏期、空间分布等多种特征应该综合运用。对于未来 SSVEP 的应用前景和方向进行了分析，SSVEP 在认知神经科学上将会有更广泛的应用，为新颖而重要的学术研究提供了机会。同时，SSVEP 的研究还有一些基本问题并未解决（例如 SSVEP 的生理学产生和传播机制尚未解决），需要我们更深入的研究。但是 SSVEP 作为 BCI 的一种输入信号，具有高信息传输率、短训练时间、实验范式简单明了及无损伤性等优点，而且在线识别正确率很高，未来将会有更广泛的应用前景和应用价值。

参考文献

[1] Hovagim Bakardjian, Toshihisa Tanaka, Andrzej Cichocki. Brain control of robotic arm using affective steady-state visual evoked potentials. Proceedings of the Fifth IASTED international conference Human-Computer Interaction(HCI 2010), 2010, 264 - 270

[2] Thorsten Lüth, Darko Ojdanic, Ola Friman, Oliver Prenzel, Axel Gräser. Low Level Control in a Semi-autonomous Rehabilitation Robotic System via a Brain-Computer Interface. 10th International Conference on Rehabilitation Robotics-ICORR. 2007, 721 - 728

[3] Faller J, Leeb R, Pfurtscheller G, Scherer R. Avatar navigation in virtual and augmented reality environments using an SSVEP BCI. Proceedings of the ICABB - 2010, 2010

[4] Sandra Mara Torres Müller, Wanderley Cardoso Celeste, Teodiano Freire Bastos-Filho, Mário Sarcinelli-Filho. Brain-computer Interface Based on Visual Evoked Potentials to Command Autonomous Robotic Wheelchair. Journal of Medical and Biological Engineering, 2010, 30(6): 407 - 416

[5] Nancy Kanwishe, Josh McDermott, Marvin M. Chun. The fusiform face area: a module in human extrastriate cortex specialized for face perception. The Journal of Neuroscience, 1997, 17(11): 4302 - 4311

[6] Dan Zhang, Bo Hong, Xiaorong Gao, Shangkai Gao, Brigitte Roder. Exploring steady-state visual evoked potentials as an index for intermodal and crossmodal spatial attention. Psychophysiology, 2011, 48(5): 1 - 11

[7] Leonard J. Trejo, Roman Rosipal, Bryan Matthews. Brain-computer interfaces for 1-D and 2-D cursor control: designs using volitional control of the EEG spectrum or steady-state visual evoked potentials. IEEE Trans Neural Syst Rehabil Eng, 2006, 14(2): 225 - 229

[8] Martinez P, Bakardjian H, Cichocki A. Fully online multicommand brain-computer interface with visual neurofeedback using SSVEP paradigm. Computation Intelligence and Neuroscience, 2007

[9] Honglai Xu, Tianyi Qian, Bo Hong, Xiaorong Gao, Shangkai Gao. A Brain-Actuated Human Computer Interface for Google Search. BMEI 2009: 1 - 4

[10] Lalor EC, Kelly SP, Finucane C, Burke R, Smith R, Reilly RB, McDarby G. Steady-state VEP-based brain-computer interface control in an immersive 3D gaming environment. EURASIP Journal on Applied Signal Processing, 2005, 19: 3156 - 3164

讲座人简介

Shangkai Gao graduated from the Department of Electrical Engineering of Tsinghua University, Beijing, China, in 1970, and received the M. E. degree of biomedical engineering in 1983 in same department of Tsinghua University, Beijing, China. She has been working in Tsinghua University since 1970. She is currently a professor of the Department of Biomedical Engineering, Tsinghua University, IEEE Fellow.

Her research interests include biomedical signal and image processing, especially the study of brain-computer interface.

五、教育神经信息工程专题

教育神经工程对计算智能的需求

John Qiang Gan

(School of Computer Science and Electronic Engineering, Essex University,
UK Research Center for learning Science Southeast University, China)

摘要:在过去的几十年中,心理学和神经科学领域的研究者们为确认与学习障碍相关的神经标记做了大量的工作,然而,到目前为止,获得的结果通常都是描述性的,而且其中的一些结论还存在争议,难以应用到实际中去。我们认为,现有方法还不足以让人们获得作为可靠诊断依据的定量神经标记。在2000年初,脑机接口(BCI)研究人员也碰到过类似的困难,直到来自信号处理和机器学习领域的研究人员参与进来,着手解决 BCI 中的问题,BCI 系统的性能和可用性才有了重大突破。本报告首先介绍了 UK 的教育神经学研究进展以及当前教育神经工程研究的不足,然后探讨计算智能可以为神经科学做些什么,包括在高维空间中特征子集的选择,利用机器学习进行分类/聚类,以及寻找新的神经记号以对学习困难进行可靠的早期诊断。

关键词:教育神经学;计算智能;神经记号

The Need for Computational Intelligence
in Educational Neural Engineering

John Qiang Gan

(School of Computer Science and Electronic Engineering, Essex University,
UK Research Center for learning Science Southeast University, China)

Abstract: A lot of effort has been made over the past few decades by researchers working in psychology and neuroscience to identify neural markers for learning difficulties/disabilities, but results obtained so far are, in general, descriptive and of these some remain controversial and thus difficult to apply in practical settings. We argue that the methods currently adopted are inadequate should one want to obtain quantifiable neural markers for reliable diagnosis purposes. This situation is very similar to that experienced by the brain computer interface (BCI) research community in the early 2000's, prior to researchers from the signal processing and machine learning communities addressing BCI problems and beginning to make major breakthroughs in the applicability and performance of BCI systems. This talk will start with a brief introduction to the educational neuroscience research in the UK and the limitations in the current educational neural engineering research and continue to address what computational intelligence can contribute to educational neural science and engineering, including feature subset selection in high-dimensional space, classification/clustering via machine learning, and searching for new neural markers for reliable early detection of learning difficulties.

Keywords：Educational neural engineering, Computational intelligence, Neural mark

1 引言

在英国,有一个名叫 Kara Tointon 的女孩,在她 7 岁的时候被发现患有阅读障碍症。幸运的是,由于神经教育学研究人员对其及时的检查和干预,她如今已有 12 年的阅读年龄了,并成为了英国一位著名的演员。为此,BBC 还有专门关于她事迹的纪录片(Kara Tointon：*Don't Call Me Stupid*),意在表明早期检查和干预以及专家教育对患有阅读障碍儿童的重要性。在英国的儿童中,有 4%～8%患有阅读障碍,3.6%～6.5%患有计算失能症,3%～6%患有注意力不集中/多动症,5%～10%有反社会行为,7%有特定语言障碍等。这些数据的下限的总和已超过 25%,为此,英国政府每年要花费大量的物力和财力去解决这些问题。英国开展了一项名为智能资本和幸福的项目,并发表在 *Nature* 上。这个项目得到的结论就是:如果一个国家要足够富裕繁荣,这个国家必须学会如何在经济和社会方面利用其公民的认知资源,而早期的干预是关键。这些都说明了神经教育学对于儿童学习困难症状的早期诊断和干预的重要性,因为解决了这些问题,必然会提高英国的经济和人民的生活质量。

2　教育神经学的研究现状

英国在教育神经学对于解决儿童学习困难的问题上开展了多个项目,并取得了众多发现,主要包括如下几个方面:

(1) 存在可以揭示学习障碍的神经标记或生物标记,这些标记在早至婴儿期就已存在;

(2) 心理失常的早期检查是可能的;

(3) 通过不同类型的干预能极大提高智力资源的范围;

(4) 终生学习对于身心健康和幸福都很重要。

因此,基于神经科学知识的教育神经工程可能是对早期诊断和干预的关键。

英国重要的教育神经科学研究中心如剑桥大学和伦敦大学,以及中国的东南大学都进行了学习障碍的相关研究,研究结果证实了在人体中的确存在关于学习障碍的神经标记,包括:

(1) 具体部位的 ERPs 信号和 EEG 节律幅度和延迟的不同;

(2) 偏少偏长的不匹配负极性,比如 MMN;

(3) 在动作观测中,感觉运动区减少的 μ 能量去同步性;

(4) 额叶 EEG 偏侧化的 alpha 能量不对称,不仅仅在被测者,在患有学习障碍的人身上都可以找到标记;

(5) 其余 EEG 能量谱的不正常(delta 低于 4Hz;theta 低于 7Hz;alpha 8~12Hz),不同频段的功率谱也可以找到相应的生物标记。

教育神经科学在读、写、算以及社会认知的神经基础的理解上取得了一定的进展。比如研究表明,不同阅读者的脑地图显示的区域也不相同(图 1)。还有干预技术方面的进展,可以在http://www.dore.co.uk/这个网址找到相关信息,包括对干预技术的评价以及检验这项技术的有效性能。

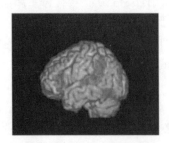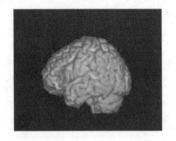

图 1　左为典型阅读者的脑地图,右为阅读障碍者的脑地图

目前教育神经科学还存在着一定的局限性,最近的进展大多是基于 EEG 信号分析,空间分辨率受到限制。而且基于 EEG 的一些神经标记主要是描述性质的,有时候很难定量评价这些神经标记,并保证它们的可靠性。比如学习障碍的神经标记就是一个描述性的假设:学习障碍与 ERPs 或者非对称的 alpha活动以及偏侧化 EEG 节律的幅度和延迟有联系。

3　计算智能在神经教育学中的应用

从模式识别的角度来说,由于复杂模式的信息不可能用几个有限的特征来包括,仅用几个特征作为复杂认知障碍的神经标记是不够的,不足以达到精确和可靠的诊断。因此,复杂模式识别的问题需要计算智能来解决。应用高等信号处理、模式识别和机器学习等计算智能的方法,我们也许可以发现和定量 EEG 和 ERPs 中与学习障碍和认知障碍相关的新特征。与心理学方法不同的是,神经科学方法是采用数据驱动的,从大量数据中获得特征。而这些数据可以被标记,也可以不被标记。

使用一些非监督的算法,可以有效地处理未标记的 EEG 数据。这个对分析儿童的数据特征非常重要,因为儿童的 EEG 数据的标记非常困难。高维的生物标记也可以通过智能计算的方法获得,而这些生物标记的神经基础可以通过基于 EEG 脑成像来研究。

寻找学习障碍中标记的特征对于学习障碍的研究是有用的。不同的特征可以从不同编码的 ERP 信号和偏侧化 EEG 的功率谱中获取。我们仅从一个实验过程中就能获得成千上万个特征,比如,功率谱密度或者带宽能量或者小波;ERP 幅度和潜伏期,AR 模型相关系数;复杂度测量,如近似熵;非高斯型、非线性特征,如双谱。但这种方式是不可行的。特征的子集数量非常大($\sum_{m=1}^{N} \frac{N!}{m!(N-m)!}$,当 N =60 时,计算量将大于 1.1×10^{18}。所以,有效的搜寻和评价方法都是非常必需的。这里说两个典型的方法,

方法一是通道选择的多目标进化方法,对每个通道进行建模,0 表示该通道未被选择到,1 表示一个被选择的通道。第一个目标是错误率的定义:$E = 1 - CV$,CV 表示 N-fold 交叉验证精度。第二个目标是已选择的通道数,目标是找到一个方法优化上述两个目标,算法主要有 Multi-Objective PSO 和 MOEA/D。举一个例子,在检测运动受伤与偏侧化中,我们从 59 个通道发现了一些有用的通道(图 2)。在运动区域中,两种方法都可以发现有用的通道,无需先验知识。不同的方法检测到不同的通道。因为 MOEA/D 方法的结果不好,图 2 显示的是 MOPSO 方法检

测到的,所以方法的选择对通道的检测很重要。因为要从上千个特征中找到一些特征,所以搜索的算法也很重要。方法二是搜索算法,FDHSFF 对搜索学习障碍的生物标记也是很有用的,流程见图 3,这是 3 个脑活动的活动地图。图 4 为心算减法的结果。

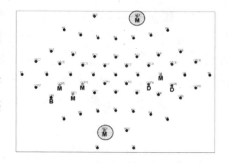

图 2　运动图像分类的通道选择

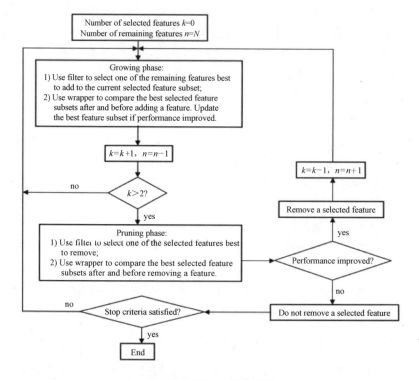

图 3　FDHSFFS 计算流程

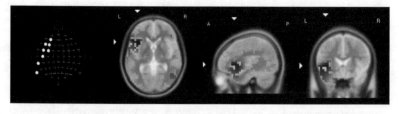

图 4　检测心算和空闲状态的通道选择以及 EEG 脑成像结果

4　总结

在 2000 年以前,关于教育神经学的研究成果相对较少,而且是描述性质的,非定量性质的。要推动教育神经科学/教育神经工程的发展,信号处理、机器学习等算法的介入非常重要。随着计算智能的驱动,将会出现更加准确可靠的诊断方法。

参考文献

[1] John Beddington, Cary L. Cooper, John Field, *et al*. The mental wealth of nations. *Nature*, 2008,455:1057-1060

讲座人简介

Professor John Q. Gan (甘强) received the BSc degree in electronic engineering from Northwestern Polytechnic University, China, in 1982, the MSc degree in automatic control and the PhD degree in biomedical electronics from Southeast University, China, in 1985 and 1991 respectively.

He used to work with Southeast University as a Professor and Head of Department of Biomedical Engineering. He is currently a Professor in School of Computer Science and Electronic Engineering, University of Essex, UK. He has co-authored a book and published over 150 research papers. He is an Associate Editor for IEEE Transactions on Systems, Man, and Cybernetics — Part B and on the editorial boards of other journals. His research interests are in biomedical signal processing, pattern recognition, data mining, machine learning, brain-computer interfaces and human-machine interaction. His home page is at http://cswww.essex.ac.uk/staff/jqgan/.

神经教育工程:教育发展的新时代

禹东川

(教育部儿童发展与学习科学重点实验室
东南大学儿童发展与学习科学中心)

摘要:神经教育学(NeuroEducation)是一个新兴的研究领域,它把认知神经科学、发展认知神经科学、教育心理学、教育学理论与技术、信息学和生物工程等领域的研究者们聚集在一起,进而研究教育和生物过程之间的相互关系。2002 年,韦钰院士在东南大学创建了我国第一个神经教育学研究中心,该研究中心主要致力于为教育实践和教育政策的制定提供理论基础。目前,研究中心已经在社会情绪能力评估方面取得了一定的成果。

关键词:教育神经科学;神经科学;教育学;社会情绪

Neuroeducation Engineering: A new Stage of Education Development

Dongchuan Yu

(Key Laboratory of Child Development and Learning Science, Ministry of Education China Research Center for Learning Science, Southeast University)

Abstract: NeuroEducation is an emerging scientific field that brings together researchers in cognitive neuroscience, developmental cognitive neuroscience, educational psychology, education theory & technology, informatics, bioengineering and other related disciplines to explore the interactions between biological processes and education. The first research center for NeuroEducation in China was created by Professor Wei Yu in Southeast University in 2002. We expected that our research in this center can provide theoretical foundation for educational practice and educational policy making. We have got some achievements on assessment of society-emotion competence.

Keywords：NeuroEducation，Neuroscience，Education，Society-emotion

1 引言

神经科学取得的一系列重大进步为教育学科的发展提供了一些重要的理论依据。神经科学研究证实，在婴儿成长过程中的某些敏感时期对特定类型的学习能力产生影响。神经科学研究同时证实，在儿童的学习过程中注意力的集中程度是非常重要的。神经科学可以提供给研究人员一个特殊的研究视角，研究人员通过它可以更好地研究儿童某些技能的学习和发展过程中的规律。

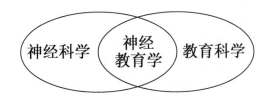

图 1　神经教育学和神经科学与教育科学的关系

回到本文讨论的话题，什么是神经教育学呢？ 基于过去十年的研究和实践，我们认为：神经教育学是一个跨学科的研究领域，它主要综合了神经科学和教育科学两门学科的内容；神经教育学也是一个交叉性的研究领域，需要把认知神经科学、实验认知神经科学、教育心理学、教育学理论与技术、信息学和生物工程等领域的研究者们聚集在一起，共同研究教育活动和神经生理过程之间的相互关系。

神经教育学以科学研究为基本态度，以实验证据为理论基础，以解决问题为研究目标，它为人的学习活动和学生教育事业提供了一个新的跨学科的理解角度。神经教育学的研究成果可以为教育实践和教育政策的制定提供必要的理论基础，它是教育科学发展的新阶段。

目前，国际上已有的神经教育学研究机构或相关机构主要有：经济合作与发展组织（OECD），智力、大脑和教育国际组织（MBE Society），神经教育学中心，国际教育学数据采集组织等。

关于神经教育学，2000 年诺贝尔生理学或医学奖得主 Eric Kandel 曾说过：如果你们愿意的话，这将是一个极好机遇：（它）可以把教育学引向一个全新的方向。

2　东南大学的神经教育学研究

2002 年,中国工程院韦钰院士在东南大学创立了国内第一个开展神经教育学研究的研究中心:学习科学研究中心。该中心的目标是:希望神经教育学研究能够为教育实践和教育政策的制定提供理论基础。中心设有儿童发展与学习科学教育部重点实验室,实验室将现代计算机信息技术、生物测量技术、情感计算方法等运用于儿童发展的行为学研究;将生物信息学和生物芯片技术运用于与儿童正常、异常发育相关的遗传学研究;将计算机图像分析、可视化算法等技术,运用于儿童发展脑机制的研究;将互联网技术、虚拟现实技术等现代信息技术,运用于儿童科学的教育研究和培训。实验室已经承担和完成了一批国家级的科研项目,在国内外学术刊物上发表了一批有影响的论文,获得了多项国家发明专利。

东南大学学习科学研究中心的研究思路是(图 2):在学习过程中,通过对某个个体或者大量个体的生理参数进行定量记录,并经过滤波、特征提取以及分类等数学方法的处理,从而实现对学习活动的定量化评价,评价结果将反馈给一些学习辅助设备,这些学习辅助设备将通过产生不同类型的刺激信号来干预或纠正学习活动,最终达到改善儿童学习活动的目的。

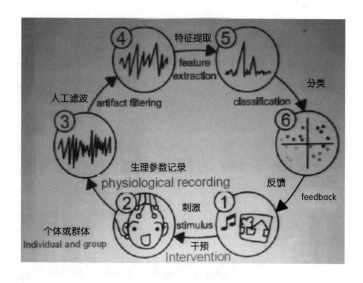

图 2　学习科学研究思路

　　学习科学研究中心在研究过程中常用的分析方法有模式识别技术、面部表情识别技术、功能性网络技术以及非线性动力学和动态网络技术。

　　进入 21 世纪，竞争越来越激烈，整合能力、协作能力、交流技巧越来越被人们所重视，而社会情绪的学习对于这些能力的培养至关重要，所以东南大学学习科学研究中心认为，儿童社会情绪的学习对未来的教育质量是决定性的，因此研究中心在儿童社会情绪能力评估方面做了大量的工作。研究人员首先通过功能磁共振成像（fMRI）和脑电图（EEG）技术来确定与社会情绪有关的大脑区域的功能结构，从而了解与社会情绪有关的神经系统的基本规律。其次，研究人员还通过面部表情识别、模式识别和复杂网络等工具来量化分析包括生理参数在内的相关数据，从而对人的情感状态进行评估。研究中心同时还开发出了用于评估社会情绪能力的心理量表以及一些相关的评估装置。此外，中心还建立了儿童成长数据库。

　　研究中心希望通过分析多种生理参数和相关数据能够确定和量化学习活动。研究人员可以测量被试者手上的体温、血氧饱和度、脉搏、皮肤阻抗等多种生理参数。图 3 展示了被试者正在进行生理参数的测量，图中右边的被试穿戴了大量的生理学参数传感器，其中也包括了采集 EEG 信号的电极。基于EEG 的学习活动研究所采用的实验范式是：让学生处于不同的学习活动中，相应地记录他们的脑电信号，然后研究学生在不同学习活动中脑电信号的变化。

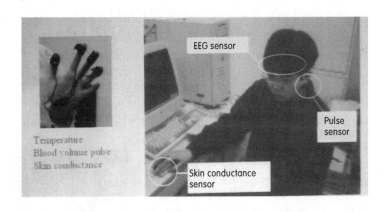

图 3　测量多种生理参数

　　研究中心还在基于图像处理的情绪识别方面做了一些尝试。图 4 展示的是研究中心开发的情绪识别系统，该系统通过获取被试者面部表情的特征向量，然后通过支持向量机（SVM）方法进行分类，从而达到识别不同类型情绪的目的。

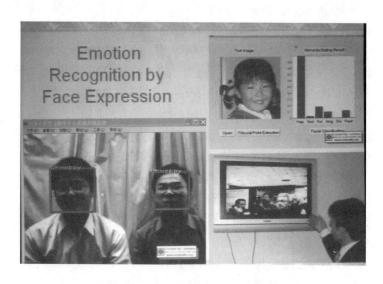

图 4　通过面部表情来识别情绪类型

3　东南大学神经教育学研究总结与展望

东南大学在神经教育学方面的有关研究为国家教育中长期发展规划的制定提供了参考意见,并得到了陈至立同志和顾秀莲同志的高度评价。研究中心承担了国家级"国培计划",为国家早期教育政策的制定提供了帮助。相关研究成果获得了教育部基础教育课程改革教学研究成果一等奖。

未来中心将与上海有关学校进行合作,力争建立国内首个神经教育学研究院,以促进相关研究的快速发展。另外,中心将开发更多的教育学设备,并用它们来进一步评估儿童的社会情绪能力,然后干预并提高儿童的社会情绪能力。此外,这些设备也可能用来进行人才的选拔。

讲座人简介

Dr. Dongchuan Yu received the M. S. degree from Southeast University in 2003 and the Ph. D. degree from Tianjin University in 2007. From March 2006 to June 2006, he was a Research Assistant with the City University of Hong Kong. From March 2007 to July

2007, he was a Guest Professor with the University of Potsdam, Potsdam, Germany. From August 2007 to October 2007, he was a Guest Expert with the National Central University, Taiwan. From December 2007 to October 2008, he was a Postdoctoral Research Fellow with the Ecole Polytechnique Federale de Lausanne (EPFL), Lausanne, Switzerland. He also visited briefly the Goettingen University, Catania University (Italy), and Weizmann Institute of Science (Israel). He is currently a Professor with the Key Laboratory of Child Development and Learning Science (Southeast University), Ministry of Education, China. He is the PI and Co-PI of several national projects and the author of over 40 published scientific papers. His research interests include nonlinear dynamics and network analysis, as well as their applications to child development and neruoeducation.

教育神经工程中的社会情绪能力评价

杨元魁

（儿童发展和学习科学教育部重点实验室，学习科学研究中心，东南大学）

摘要：社会情绪能力是心理学和教育学的研究热点，但由于缺乏相应的科学实证，目前仍然很难对社会情绪能力进行评价。本报告首先介绍社会情绪能力评价框架的概念和神经工程中各种不同技术工具的应用，讨论各种可用于评估社会情绪能力以及社交躲避行为的方法；然后以社会情绪能力中最重要的换位思考能力为例，说明各种不同评估方法的作用；最后，结合我们的工作，对社会情绪能力、实证道德教育以及儿童社交躲避行为的将来工作进行展望。

关键词：社会情绪能力；评价框架；儿童教育

Assessment of Social Emotional Competence in Educational Neural Engineering

Yuankui Yang

(Key laboratory of Child Development and Learning Science, Ministry of Education, Research Center of Learning Science, Southeast University)

Abstract: Social emotional competence has become the research hotspot in psychology and education. However, the assessment of social emotional competence is still very difficult and lack of scientific evidence. In this talk, the concept and assessment framework of social emotional competence will be firstly introduced. Secondly, the application of neural engineering tools and methods for assessing social emotional competence and social withdraw behavior will be discussed. Thirdly, efforts on assessment of empathy — one of the most important social emotional competence — will be presented as an example. Finally, the future work on social emotional competence, evidence-based moral education and children's social withdraw behavior will be talked.

Keywords：Social emotional competence, Evaluation framework, Children's Education

1　引言

国际 MBES 协会(International Society of Mind, Brain and Education)成立于 2003 年,创始人是小泉英明教授,该协会于 2007 年开始发行同样由小泉教授创刊的 *Mind, Brain and Education* 期刊。MBES 协会的宗旨是集成多种手段对人类的学习和发展进行研究,借助教育学、生物学、认知科学等相关学科知识背景开创人类思维、大脑和教育研究的新领域。

儿童社会情绪能力是一种在社会情景中做出有效和适当行为的能力,不仅包括了儿童所取得的社会成果,比如交友、受同伴欢迎、积极参与同伴间的交往活动等,也包括了同伴、教师和他人的评价,以及发起并保持合作的实际社会行为。对个人而言,社会情绪能力包含五方面的内容,分别是了解自我的情绪——自我认同,管理自我的情绪——自我约束,鼓励自我的情绪——自我激励,识别他人的情绪——相互理解,以及处理人际关系——社交技巧。相应地,社会能力是儿童为顺利适应社会而具备的社会、情绪、认知方面的技能和行为,社会情绪能力和社会能力相互影响,儿童社会情绪能力的高低会对其日后的社会能力及学业能力产生重要的影响。

中国的独生子女政策形成了中国家庭特有的 4-2-1 结构,造成了中国独生子女强烈的自我意识,尤其在农村,留守子女群体由于无法享受到充分的教育,没有足够的学习机会,使得其社会情绪能力的发展出现了偏差,形成了如反社会、偏激、社交退缩等行为问题,对其社会能力的发展产生了重要的负面影响。

在对社交退缩行为进行研究后发现,杏仁核对新刺激的反应水平可能是造成社交退缩行为的神经学基础,同时,额叶脑电图信号的不对称性也会造成性格内向的人产生社交退缩行为,这些研究找出的生物学标记可为进一步理解社交退缩行为及相应的社会情绪能力奠定基础。

2　评价方法和工具

为对儿童的社会情绪能力进行评价,作者所在研究机构设计建立了一个评价框架,并自制了所需的相应设备。该评价框架集成了三种不同的评价方法,分别是传统的心理学评价方法、行为学评价方法和生理学指标评价方法。其中心理学评价方法包括了问卷调查、大规模群体调查以及统计调查方法。行为学评价方法包括行为编码、表情分析、语音识别、语言分析和观察评鉴指标,而这其中如表情分析、语音识别、语言分析等都需要大量计算机相关知识的协助。生理学指标评价方法包括功能磁共振成像、近红外成像分析、脑电图/事件相关电位分析、心电图、皮肤电导性、体温、血压、呼吸、眼动分析等生理参数和指标。为测定相关生理参数和指标,东南大学还自制了两套用于测定生理状态的无线电生理测量设备,其一为可用于测定心率、皮肤电导性和体温的无线可穿戴式设备;其二为可用于测定心率及心率变化情况的头戴式设备,之所以采用头戴式设备,是因为考虑到戴上该设备后孩子们还可以自由参加各种活动,不会对他们的行为产生任何影响,同时结合无线模块,还可构成无线传感器网络,可同时对 24 个孩子的心率及变化情况进行测定。

3　对换位思考能力的评价

以儿童在探究性科学教育(Inquiry-Based Science Education,IBSE)过程中表现出的换位思考能力为例,作者结合上述评价工具和框架,对孩子们的这种能力进行了评价,分别采用了传统心理学评价方法中的规模调查和统计调查方法,行为学评价方法中的行为编码和观察评鉴指标,以及生理学指标评价方法中的情绪激发和皮肤电导性参数方法。

探究性科学教育过程是让孩子们在课堂上通过听老师讲课和相互合作的方式进行学习的过程,强调的是学生自我构建学习过程,而不是被动地接受老师的灌输。在这个过程中,认真听课和相互合作是两项很重要的学习技能,这两项技能也是研究中所要关注的。在同学间的相互合作中,经常会出现这样的问题,即有些男同学由于害怕女同学觉得自己太过活跃、太过主动影响小组的团队合作而无法参加到小组的合作实验活动中去,进而对该同学的情绪造成不良影响。这也是同学缺乏换位思考能力的体现。而探究性科学教育方法可有效地提高同学的换位思考能力,这也是该项研究所要进行的工作。

4　实验结果

在对幼儿园和小学进行长达七年的研究之后,作者用包括规模问卷调查、观察评鉴指标以及行为学编码和皮肤电导性测量结果对接受探究性科学教育过程后学生的换位思考能力进行评价。规模调查问卷的结果显示,在对小学一年级的课堂进行一年的跟踪研究后,IBSE 科学教育方法可有效地提高同学的换位思考能力;利用观察评鉴指标,可将学生的行为分为三个层次,利于老师和研究人员对同学的行为表现进行打分,结果表明,接受 IBSE 教育方法的同学得分要明显高于没有接受 IBSE 教育方法的同学。同样地,对教室中同学间的相互协作能力进行行为学编码,结果表明,随着时间的推移,与没有接受探究性科学教育过程的同学相比,他们表现出了更为明显的协作性,而对皮肤电导性进行测量的结果表明,具备更强换位思考能力、更具协作性的同学的皮肤电阻值会下降得更快。这些结果都充分说明了 IBSE 科学教育方法的有效性,也表明了这种评价框架和方法的实际作用。

5　未来展望

以上实验结果已经说明了这种评价框架和评价方法的有效性和适用性。但现有的工作仅仅是建立了一个评价框架,集成了若干种现有的评价方法,进行数据的采集和比对。将来的工作主要面向实证道德教育、社交退缩行为的研究,可以构建一个评价—界定—干预的研究回路,对社交退缩行为开展深入地研究,而目前的工作仅是一个开始。

报告人简介

Dr. Yang is a lecturer of Key Laboratory of Child Development and Learning Science (Southeast University). He has a background of Biomedical Engineering, Developmental Psychology, and Science Education. His research interests include children's social emotional competence, evidence-based moral education, children's social withdraw behavior, etc.

智能机械手 **SmartHand** 的设计及实验评估

Maria Chiara Carrozza

(Scuola Superiore Sant'Anna (SSSA)，Pisa，Italy)

摘要:手在日常生活中是非常重要的,由于疾病或事故失去手将会引起严重的生理和心理障碍。虽然失去手对于一个人而言有很严重的影响,但由于对假手的需求太小以至于在过去的 40 年里没有引起重视,在控制接口和机械设计方面都没有重大的改进。我们研究并设计了一种拟人式的机械手,称为 SmartHand,它由 5 个手指组成,通过 4 个电机驱动,具有 16 个自由度。手的设计和开发采用了欠驱动的手指和差动机构,从而使它能适应不同大小的物体。SmartHand 可以实现日常的抓握、数数和用食指指向目标。它同时也集成了 40 个感受本体和外部信息的传感器,可以实现自动控制,以及通过特定的传入接口为截肢者提供感觉反馈。借助于自行设计的嵌入式控制器,它能够执行控制环路并能与外界环境进行双向的信息交互。机械手的重量和真手的重量相近,与一些商用的假手也相差无几(机械手重 530g,外加 145g 的腕)。SmartHand 的速度与商用假手类似,完成一次开合只需要 1.5s,能够稳定地抓握 3.6kg 的物体,提起 10kg 的箱子。

关键词:假肢;灵巧手;人机接口;嵌入式控制;感觉反馈

On the Design and Experimental Evaluation of the SmartHand

Maria Chiara Carrozza

(Scuola Superiore Sant'Anna (SSSA)，Pisa，Italy)

Abstract: The hand is a powerful tool and its loss causes a severe psychological and physical drawback. Despite the significant impact of losing a hand，numbers of amputees requiring prosthesis are too small to push manufacturers to innovate their products，so that both control interfaces and mechanisms have almost not changed in the past 40 years. A new anthropomorphic transradial prosthesis named SmartHand has been designed and developed. It is a five fingered self-contained robotic hand，with 16 degrees of freedom，actuated by 4 motors. Underactuated fingers and differential mechanisms have been designed and exploited in order to fit all functional components in the size and weight of a natural human hand. The hand is able to perform everyday grasps, count and independently point the index. It integrates 40 proprioceptive and exteroceptive sensors both employed for implementing automatic control and for delivering sensory feedback to the amputee，by means of suitable afferent interfaces. This is possible due to a customized embedded controller，able to execute control loops and bi-directionally exchange information with the external world. The weight of the hand is similar to the natural hand weight and comparable to actual commercial prostheses（hand 530g plus 145g for the standard wrist attachment）. Speed is comparable to commercial prostheses：the hand fully opens/closes in 1.5 seconds，being able to stably grasp up to 3.6 kg objects and lift a 10 kg suitcase.

Keywords：Prosthesis；SmartHand；Human-machine interface；Embedded control；Sensory feedback

1 引言

在过去的 40 年里,机械手的设计及应用没有能够引起重视,在控制接口和机械设计方面都没有重大的进步。然而手的功能在日常生活中非常重要,失去手将会引起一系列生理的和心理的障碍[1]。

我们团队的研究目标是设计和研发灵巧的机械假手,并可以通过大脑或其他神经接口来控制,为残疾人提供基于神经控制的假手(Prosthetic Hand)。本研究的重点包括机械手的底层控制(Low-Level Control)、机械手对人体的反馈以及一些双向的非植入式的人机接口。机械手的机械设计及人机接口研究是其中的关键。在神经假手中所遇到的主要挑战来自三个方面：

(1) 能实现灵巧抓握的机械手的设计；

(2) 能够控制机械手的人机接口；

（3）机械手到人的反馈的实现。

2　SmartHand 的机械设计

SmartHand 是我们最新设计的一种灵巧手,它由 5 个手指组成,大小与真实的人手等同,它由 4 个电机控制,分别用来控制大拇指的弯曲、食指的弯曲、中指无名指小拇指的弯曲以及大拇指的内收和外展。不同的组合可以实现不同的抓握,如握、捏、侧抓等[2]。欠驱动的手指及差动机构被应用于手的设计,从而使它可以适应不同大小的物体。这种机械手可以进行日常的抓握、数数和用食指指目标。它同时也集成了 40 个感受本体和外界信息的传感器,用于实现自动的控制以及通过特定的传入接口为截肢者提供感觉反馈。借助于自行设计的嵌入式控制器,它能够执行控制环路并能与外界环境进行双向的信息交互[3]。机械手的重量和真手的重量差不多,与一些商用的假手相差无几(机械手重 530g,外加 145g 的腕),机械手完成一次开合只需要 1.5s,能够稳定地抓握 3.6kg 的物体,提起 10kg 的箱子[4]。SmartHand 的机械结构、抓握功能以及在残疾人士的使用情况如图 1 所示。

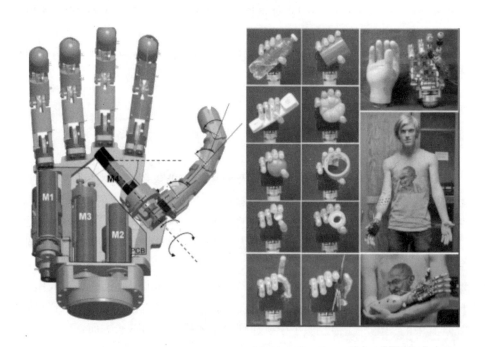

图 1　SmartHand 的机械结构、抓握功能及用于残疾人的展示

3 机械手的控制接口研究

3.1 基于肌电(EMG)的实时机械手控制[5]

我们的研究表明在有视觉反馈的条件下,用8通道的表面肌电(如图2所示)来实时连续控制7种机械手的手指运动;实验分别在5个截肢的病人和5个正常人身上进行,实验结果显示,这种利用肌电对机械手控制的方法简单可行,能够完成7个不同姿势的手指动作(如图2所示),可以用于实际的应用。其中,对于正常人可达到89%的正确率,对于截肢的残疾人也可以达到79%。

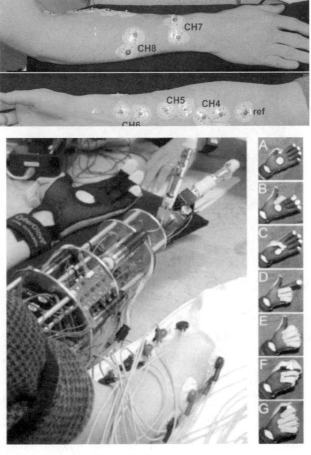

图2 肌电位置示意图(上)及系统装配图(下)

3.2　基于主成分分析的机械手控制[6]

在传统的肌电控制多自由度机械手中,通常需要使用者去调节不同通道的肌电来使机械手独立地运动,是一个多对多的映射。这里,我们运用基于独立主成分分析(PCA)的方法用两通道的输入来控制一个 16 自由度的机械手完成 50 种不同的抓握动作(图 3)。实验中用鼠标的二维坐标值来代替 2 通道的肌电,结果验证了基于 PCA 的机械手控制方法可以用于实现多种不同的手的形状和抓握。

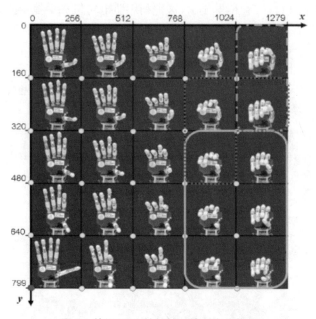

图 3　基于 PCA 的机械手控制方式模拟

3.3　基于智能视觉系统的机械手控制[7]

我们开发了一种新的基于摄像头的控制系统,该系统由装在机械手上的摄像头、激光器和距离传感器等三部分组成。控制系统利用摄像头识别出激光器瞄准的目标物体,再综合由距离传感器获得的目标物体距离,计算出目标物体的大小,然后控制机械手的抓握形状,最后通过肌电触发来实现对目标物体的抓握(图 4)。该系统经过 18 个被试者的测试,结果显示准确识别目标物体的正确率可达 84%,另外研究同时发现有 6% 的情况虽然被试者不能给出最优的抓握方式,但也可以完成抓握的任务。研究结果意味着该系统在临床的应用中将有非常好的前景。

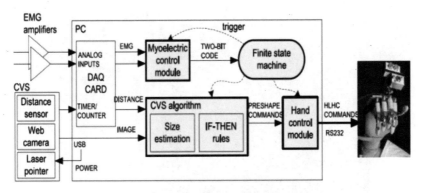

图 4 系统整体框图的实现

3.4 基于外周神经系统(PNS)修复的控制[8]

我们还进行了外周神经系统(PNS)修复的研究,我们用 tf-LIFE 电极植入截肢患者的手臂,采集外周神经信号并进行机械手的控制。结果表明,从 PNS 中至少可以提取出 3 种不同的抓握信息,并用于机械手的控制。同时,通过对 PNS 进行电刺激还可以为截肢患者提供感觉反馈。

4 反馈研究及与手的集成

4.1 振动的反馈对人体的效果作用评价[9]

我们研究了人体对振动的反馈效果作用。首先,我们开发了一个振动反馈系统,并将该系统应用于人的振动反馈区分实验中,研究不同频率、不同幅度、不同距离、不同朝向的振动在人体上能否被区分出来。实验结果表明,在不考虑刺激频率的情况下,至少可以区分 3 种不同的幅度,也可以区分不同的频率;刺激的最小区分距离为 1.5cm,若小于这个距离则人体不能进行分辨。

4.2 带触摸功能的机械手指设计[10]

我们设计了一种带触摸功能的机械手指,这种手指装有 MEMS 工艺生产的指纹传感器(如图 5 所示),表面覆盖带指纹的人工皮肤。用该手指去触摸不同粗糙度的物体时,我们记录传感器的信号,结果表明该机械手指可以区分不同的粗糙度。该手指有望集成到现有的机械手中。

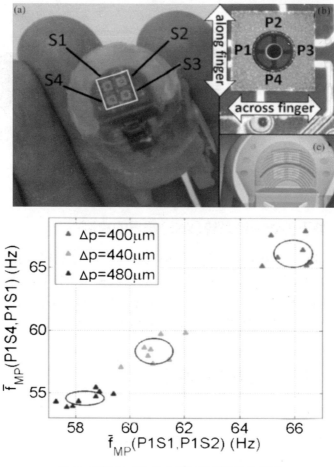

图 5　手指 MEMS 传感器及手指区分效果图

4　总结与展望

那么,我们下一步的研究目标是什么呢? 目前,我们的研究重点集中在机械手的测试、技术的改进和多学科的交叉发展。我们现在需要开发更好的人机接口来完成神经假肢的目的。我们主要关注两种设计方式,一种是用户驱动(user-driven)的设计方式,该方式关注用户的负担、需要的注意力和给用户的反馈量,目前的研究表明用户对于假肢系统的接受程度取决于所需要的注意力,而不是单单的抓握成功率。另一种是技术驱动(technology-driven)的设计方式,它使我们的手能够适用于不同接口系统来完成各种任务。

参考文献

[1] Cipriani C, *et al*. On the shared control of an EMG-controlled prosthetic hand. Analysis of user-prosthesis interaction. *IEEE Transactions on Robotics*, 2008, 24: 170 – 184

[2] Cipriani C, Controzzi M and Carrozza MC. Objectives, criteria and methods for the design of the SmartHand transradial prosthesis. *Robotica*, 2010, 28(6): 919 – 927

[3] Cipriani C, *et al*. Embedded hardware architecture based on microcontrollers for the action and perception of a transradial prosthesis. In: Biomedical Robotics and Biomechatronics, 2008. BioRob 2008. 2nd IEEE RAS & EMBS International Conference on. 2008

[4] Christian Cipriani, MCAM. The SmartHand transradial prosthesis. *Journal of NeuroEngineering and Rehabilitation*, 2011. doi: 10. 1186/1743 – 0003 – 8 – 29

[5] Cipriani C, *et al*. Online Myoelectric Control of a Dexterous Hand Prosthesis by Transradial Amputees. *IEEE Trans Neural Syst Rehabil Eng*, 2011

[6] Matrone GC, *et al*. Principal components analysis based control of a multi-DoF underactuated prosthetic hand. *J Neuroeng Rehabil*, 2010, 7: 16

[7] Dosen S, *et al*. Cognitive vision system for control of dexterous prosthetic hands: experimental evaluation. *J Neuroeng Rehabil*, 2010, 7: 42

[8] Micera S, *et al*. Decoding Information From Neural Signals Recorded Using Intraneural Electrodes: Toward the Development of a Neurocontrolled Hand *Prosthesis*. *Proceedings of the IEEE*, 2010, 98(3): 407 – 417

[9] D'Alonzo M, Cipriani C, Carrozza MC. Vibrotactile Sensory Substitution in Multi-fingered Hand Prostheses: Evaluation Studies. In: Proc. of IEEE RAS/EMBS Intl. Conf. on Rehabilitation Robotics, ICORR, Zurich, Switzerland, Jun. 29 – July 1, 2011

[10] Oddo CM, Controzzi M, Beccai L, Cipriani C, Carrozza MC. Roughness Encoding for Discrimination of Surfaces in Artificial Active Touch. *IEEE Transactions on Robotics*. In press, 2011

讲座人简介

Prof. Maria Chiara Carrozza received the Laurea degree in physics from the University of Pisa, Italy, in 1990 and the PhD in Engineering at Scuola Superiore Sant'Anna (SSSA), in 1994. Since November 2006, she is Full Professor of Biomedical Engineering and Robotics at Scuola Superiore Sant'Anna. Since Nov. 2004 to Oct. 2007, she was Director of the Research Division and elected Member of the national Board of the Italian association of Biomedical Engineering. Since Nov. 2007, she is Director of Scuola Superiore Sant'Anna. She was visiting professor at the Technical University of Wien, Austria, with a graduate course entitled Biomechatronics, she is involved in the scientific management of the Italy-Japan joint laboratory for Humanoid Robotics ROBOCASA, WasedaUniversity, Tokyo, and she is Guest Professor at the Zhejiang University, Hangzhou, China. She has scientific and coordination responsibilities within several research projects, funded under the Sixth and Seventh Framework Programme of the European Union (some recent projects are Nanobiotact, Nanobiotouch, Evryon, Smarthand, Neurobotics, Robotcub, Cyberhand) and under national and regional programmes (some recent projects are OpenHand, Tectum, RITA, Safehand, Neuro-Bike, Enable). Since 2004 to 2007, she was the Coordinator of the ARTS Lab of SSSA. In the period 2006 – 2009 she supervised more than 30 PhD, Master and Bachelor theses and she currently leads a group of about 35 researchers, PhD students and research assistants. She is author of several scientific papers (more than 60 ISI papers and more than 100 papers in referred conference proceedings) and of 12 national and international patents. She served as Editor of several Special Issues of International Journals, as Member of Committees for International Conference organization and she gave several invited lectures and plenary speeches to national and international conferences and she is a recipient or numerous awards. She is member and of the IEEE Robotics and Automation Society (RAS) and of the IEEE Engineering in Medicine and Biology. She is member of the RAS Technical Committee "Micro/Nano Robotics and Automation". Her research interests are in ambient assisted living, technical aids, biorobotics, rehabilitation engineering, bionics, cybernetic hands, humanoid robotics, systems for functional replacements and augmentation, biomechatronic interfaces, tactile sensors, artificial skin, applications of renewable energy to robotics and of robotics to renewable energy.

绕过损伤的脊髓:用皮层控制的功能电刺激实现上肢的抓取功能

Lee Miller

(Department of Biomedical Engineering, Northwestern University,
Evanston, Illinois, United States of America)

摘要:颈中部的脊索损伤会严重降低人的自主生活能力,与此同时它还会导致肌肉张力和骨密度降低,或引起心血管系统功能的衰弱。功能电刺激可以刺激支配肌肉活动的神经,使瘫痪的肌肉重新产生动作。目前功能电刺激手段已经被用于数百个病人帮助他们完成基本的自主抓握动作,另外还被用于手的伸抓、膀胱和肠胃控制、姿态和运动的辅助,以及间接地帮助改善脊索损伤患者的健康状况。

目前控制这类抓握动作的都是事先制定好的刺激模式,而刺激的触发也仅仅是病人残余肢体动作中提取出的简单信号,这就把手部功能限制在少数几个事先制定好的模式。而我们开发的系统使用猴子运动皮层中记录到的神经信号作为控制信号,为自主控制多个肌肉完成更多任务提供了可能。为实现这一目标,我们必须提取出动态的运动信息,特别是神经元上记录到的自主肌肉活动相关的信息。这与其他脑机接口应用中仅仅依赖动作信息有很大的不同。

开发这样的一个系统并非易事。在特定的实验室条件下,我们可以计算出运动皮层活动与正常运动时相关肌肉活动间的传递函数。我们做了线性和非线性的解码,将记录到的神经元信号转化成用于多通道刺激器的控制指令。我们已经在多个猴子上使用了这些方法,使外周神经被阻断后暂时瘫痪的猴子重新恢复了自主控制腕部等长收缩以及简单抓握动作的能力。目前我们正致力于提高控制的自由度。在临床条件下,我们无法得到正常系统的输出,因此控制器必须能自我校正以适应神经信号的丢失或改变,但是这种变化又不应该干扰用户使之改变之前的习惯。最后,系统还应该能在不同的动作和姿势下不需要校正就能产生输出。如果我们能够使用猴子模型在实验室条件下解决这些问题,我们相信类似的脑部控制的功能电刺激假体可以最终使因高位截瘫丧失控制手部动作能力的病人受益。

关键词:脑部控制;FES;手部功能恢复

Bypassing the Injured Spinal Cord: A Demonstration of Cortically Controlled Functional Electrical Stimulation for Grasp

Lee Miller

(Department of Biomedical Engineering, Northwestern University, Evanston, Illinois, United States of America)

Abstract: Spinal cord injury at the mid-cervical level causes a devastating loss of independence, as well as a host of adverse physiological changes including loss of muscle tone and bone density, and compromised cardiovascular system. Functional Electrical Stimulation (FES) can be used to produce movement in paralyzed muscles by the application of electrical stimuli to the nerves innervating the muscles. FES of forearm and hand muscles has been used to provide basic, voluntary hand grasp to hundreds of human patients, as well as reaching, grasp, bowel and bladder control, stance, and locomotion. Indirectly, FES also can significantly improve the overall health of spinal injured patients through the exercise it generates.

Current approaches to the control of grasping movements use pre-programmed patterns of muscle activation that are controlled by the patient using simple signals derived from residual movement of the proximal limb. However, this limits the available hand function to the few tasks programmed into the controller. In contrast, we are developing a system that uses neural signals recorded from a multi-electrode array implanted in the motor cortex of monkeys; this system has the potential to provide independent control of multiple muscles over a broad range of functional tasks. To realize this potential, we must extract information about movement dynamics, specifically intended muscle activity from the neural recordings. This goal runs counter to that of virtually all other brain machine interface applications that rely exclusively on kinematic information.

The challenges of such a system are considerable. In the laboratory setting we can compute transfer functions between motor cortex activity and the corresponding muscle activity during normal movements. We are developing both linear and nonlinear decodes that transform neural recordings into multi-channel stimulator pulse-width commands. We have used these methods to allow several monkeys to regain voluntary control of isometric wrist torque and simple grasping movements despite temporary paralysis induced by peripheral nerve block. We are currently working to increase the number of independently controlled degrees of freedom. In a clinical setting, this must all be accomplished without initial access to the normal system outputs. The controller must be readily recalibrated to accommodate lost or altered neural signals, but these changes must not interfere with parallel adaptive changes made by the user. Finally, the system must be able to generalize across a range of different dynamical and postural settings without recalibration. If we are able to solve these problems using a monkey model in the laboratory setting, we believe that similar brain-controlled FES prostheses might ultimately benefit patients with higher cervical injuries who lack even the ability to control proximal arm movements.

Keywords：Brain control，FES，Hand function recovery

1　引言

功能电刺激在康复领域里应用广泛，比较常见的应用有足下垂刺激和呼吸辅助刺激。前者通过电刺激对患者步态进行矫正，后者通过刺激呼吸肌来辅助患者呼吸。在西北大学与凯撒西储大学的合作项目中，功能电刺激被用于手部功能的恢复。

在美国，每年有约 11000 例新的脊索损伤病例，而整个美国总共大概有250000 人生活在脊索损伤的痛苦中。脊索损伤对病人的生活带来极大的不便，而临床上对这种损伤没有有效的治疗手段。因此有关这类病人的康复治疗研究对改善这些病人的生活极为关键。运用功能电刺激的手段可以帮助恢复脊索损伤患者部分功能，比如有一个属于 C5 损伤的病人，因为损伤手臂已经不能自主抓握，不过该病人仍旧可以移动肩膀。研究人员通过他的肩膀动作来促发电刺激帮助其实现抓握。目前这个病人使用的康复装置已经被用在

200～300个病人身上。在该装置的帮助下，病人可以成功抓起物体，甚至能拿起笔并进行书写。这种装置能极大地恢复病人的手部功能，但是由于该装置中所有的刺激模式都是事先制定好的，肩部动作只是作为一个触发器来触发刺激，故它的功能十分有限。为了解决肩膀动作只能触发一个或少数几个动作的问题，我们在研究中将脑部信号代替肩部动作作为控制信号。

2 实验方法

我们实验室采用猴子作为实验对象，并将猴子脑部信号采集出来并转化成声音信号。我们发现猴子伸出手时声音变得尖锐，这说明手部动作跟我们记录到的脑部信号是有一定的相关性的。与此同时，我们记录了猴子手部不同肌肉的 EMG 信号，发现这些信号是不一样的，与此同时不同的神经元中记录到的信号也不相同。由于不同的神经元有关于不同肌肉的信息，因此我们结合这两个信号，同步记录和比较后可以重建特定动作所涉及的所有肌肉活动。

我们在 PMD、M1 区各植入 100 通道的犹他电极，然后训练猴子完成特定动作。另外，记录 EMG 信号的电极是犹他斜电极（如图 1 所示），用于功能电刺激的是卡夫电极（如图 2 所示）。

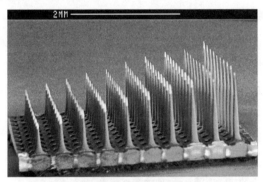

图 1 犹他斜电极

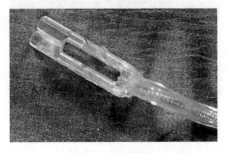

图 2 卡夫电极

电极被埋在手部的三根主要神经上,电极埋设的位置如图 3 所示。

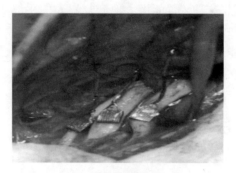

图 3　手部电极埋设位置

在第一个实验中我们训练猴子利用手腕的屈伸来控制电脑屏幕上光标的移动。我们想通过这个实验研究记录到的脑部信号跟 EMG 信号以及手腕动作三者之间的特定关系,解码神经信号。在第二个实验中我们用利多卡因局部麻醉了猴子手部,此时猴子的自主腕部动作已经消失,然后利用 FES(功能电刺激)的手段来帮助猴子恢复腕部屈伸功能。在该实验中我们想检测神经解码结果控制的 FES(功能电刺激)用于恢复手部功能的效果。

3　实验结果

图 4 显示了猴子进行一维腕部屈伸实验时同步记录到的各个信号。我们可以看到神经元的兴奋与腕部动作同步,而实际的 EMG 信号与预测的 EMG 信号吻合得很好。

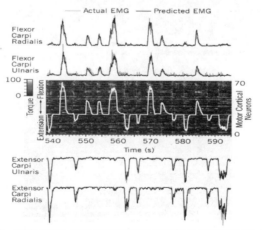

图　实际 EMG 信号与神经解码后重构的 EMG 信号对比

在图 5 中可以看到我们施加的 FES 刺激是否达到了预期效果。我们根据记录到的神经元信号重建出 EMG 信号，然后根据该 EMG 信号在特定肌肉上施加 FES，在图中灰色矩形部分表示猴子没能完成腕部屈伸实验，而白色矩形部分表示猴子成功完成了腕部屈伸实验。从时间轴上可以看出在关掉 FES 的同时猴子不能完成腕部实验，但是仍旧可以看到有相应的神经活动，这说明虽然猴子没能完成实验，但是有意愿要做出实验中的动作。而白色矩形部分发生在 FES 开启的时间端，说明我们的 FES 是切实有效的，能帮助猴子完成实验。

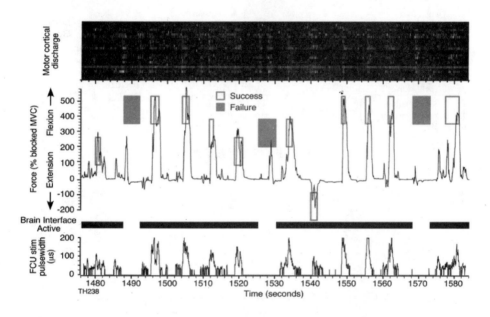

图 5　受脑部控制的 FES 信号和相应刺激产生的腕部力量

除了上述的腕部一维解码实验外，我们还做了猴子的抓握实验。图 6 是本次抓握实验的系统示意图。这个实验中猴子同样被注射利多卡因，局部麻醉猴子手部动作。麻醉后的猴子仍旧可以张开手指，也能够伸出手臂，但是不能抓握。在实验过程中猴子将手放在触摸板上，然后伸出手去抓球，并将球放到管子里。该装置中有两个红外线传感器，可以检测到球被抓起和放进。

在图 7 中我们可以看到 FES 在猴子抓握实验中起的作用。图中，虚线表示使用 FES 后实验的成功率，实线表示没有使用 FES 的实验成功率，显然使用 FES 后，实验成功率变高了。

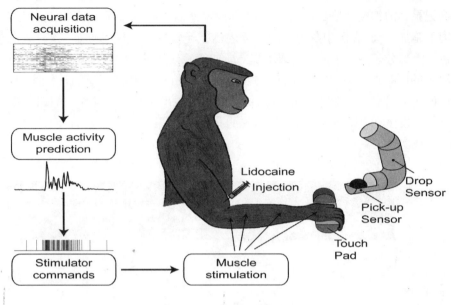

图 6　抓握实验系统示意图

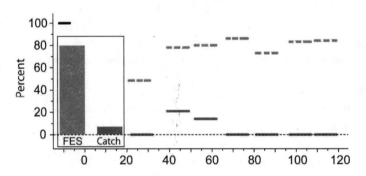

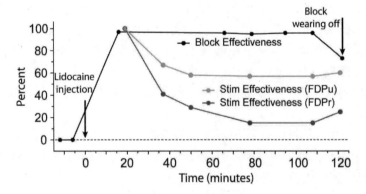

图 7　猴子抓握实验结果

与此同时使用 FES 后猴子在完成一次成功动作所耗费的时间比神经阻断后不使用 FES 耗费的时间短（具体结果见图 8）。

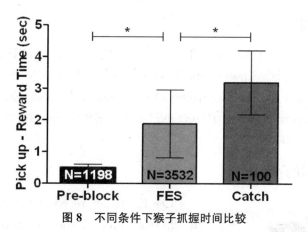

图 8　不同条件下猴子抓握时间比较

4　讨论与总结

上述一切说明，通过神经信号控制 FES 恢复瘫痪病人的手部功能是可行的。目前我们实验室跟凯撒西储大学的合作中研究了 FES 在人身上的应用。

如图 9 所示的是一个 C3、C4
损伤的病人，她只能移动下巴。研
究人员采集了她的下巴动作用作
控制 FES 输出刺激使得部分手部
动作得以恢复。

我们可以预测出肩部、手臂、
手掌等处肌肉的高质量 EMG 信
号。我们也可以直接利用大脑皮
层信号驱动多通道的刺激器用于
实现抓握。目前我们研究了多种
模式的输入信号，包括神经元信

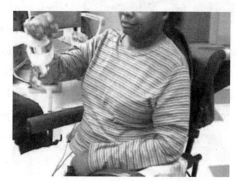

图 9　病人利用 FES 控制手部动作

号，EMG 信号以及眨眼动作。这些信号被用于 FES 驱动手臂动作的控制信号。随着研究的深入，FES 必将帮助瘫痪病人完成更多实际任务，这就对控制信号提出了很高的要求。为实现辅助病人完成多任务的目标，控制信号将会更加复杂。而我们的研究中，从脑部神经信号解码，将解码结果转化成 FES 控制指令使得辅助瘫痪患者完成多种任务成为可能。因此在未来的研究中,利用神经解码结果指导 FES 输出将是瘫痪患者康复研究中比较重要的一个部分。

参考文献

[1] Pohlmeyer EA, Oby ER, Perreault EJ, Solla SA, Kilgore KL, *et al*. (2009) Toward the Restoration of Hand Use to a Paralyzed Monkey: Brain-Controlled Functional Electrical Stimulation of Forearm Muscles. *PLoS ONE* 4(6): e5924. doi:10. 1371/journal. pone. 0005924

讲座人简介

Lee E. Miller received the B. A. degree in Physics from Goshen College, Goshen, IN, in 1980, and the M. S. degree in Biomedical Engineering and the Ph. D. degree in Physiology from Northwestern University in 1983 and 1989, respectively. He completed two years of postdoctoral training in the Department of Medical Physics, University of Nijmegen, Netherlands. He is currently a Professor in the Departments of Physiology, Physical Medicine and Rehabilitation, and Biomedical Engineering at Northwestern University. His primary research interests are the cortical control of muscle activity and limb movement, and the development of brain-machine interfaces that attempt to mimic normal physiological systems.

多功能上肢假肢的仿生控制

李光林

（中国科学院深圳先进技术研究院）

摘要：截肢患者一直期待拥有高性能义肢，以使他们恢复肢体运动功能。当前上肢截肢患者使用的人工假手并不能使患者充分恢复手与臂的运动能力。使用模式识别方法改进义肢的控制性能，从肌电信号（EMG）中提取出神经信息特征实现多自由度的运动控制。本文开发使用了一种称为目标肌肉神经分布重建（TMR）的新型神经–机器接口，利用更多的肌电信号提高假手的控制性能。

关键词：仿生控制；肌电信号；目标肌肉神经分布重建（TMR）

Bionic Control of Multifunctional Prostheses for Upper-Limb Amputees

Guanglin Li

(Shenzhen Institutes of Advanced Technology Chinese Academy of Sciences)

Abstract：Limb amputees have always been expecting high-performance artificial arms to restore the motion functions involved in their lost limbs. Currently available prostheses following upper-limb amputation do not adequately restore the function of an individual's arm and hand. To improve the control performance of artificial arms, pattern recognition has been proposed in multifunctional prosthesis control to extract neural information from electromyographic (EMG) signals, allowing intuitive control of multiple degrees of freedom. To get more EMG signals for control of myoelectric prosthesis, a new neural-machine interface, called targeted muscle reinnervation, has been developed and used for enhancing the control performance of a prosthesis.

Keyword：Bionic control，Electromyogram，Targeted muscle reinnervation（TMR）

1　引言

中国有超过 2400 万人肢体残疾，其中约有 230 万截肢患者需要义肢恢复失去手臂的肢体功能。当前，如何提高人工手的控制性能并增加肢体功能，仍然是一个巨大的挑战，特别针对残疾程度严重的高位截肢情况。

当前大部分肌电假肢通过肌电信号控制。电极检测到因肌肉收缩产生的电信号（EMG）后，给假肢发送相应的控制信号。现行方法使用电极从剩余肌肉获取表面 EMG 信号，用不同 EMG 信号分别驱动不同的动作。

现有利用肌电信号控制假肢的方法存在两个局限：① 病人截肢后，本身就缺少肌肉源。现有方法需从一对剩余肌肉获取 EMG 信号，一对 EMG 信号控制一个自由度。截肢患者可以利用的肌肉数量有限，因此很难实现多功能假肢的控制。例如，对于肘关节截肢患者，若用二头肌信号控制手的打开，用三头肌控制手的握紧，就几乎没有剩下的肌肉可供实现其他动作了。② 控制策略并不合适。手臂剩余的肌肉跟失去的肢体功能之间没有自然的联系。例如，我们用二头肌、三头肌控制肘的伸展与弯曲，若要截肢病人用二头肌、三头肌控制手的打开和握紧则非常困难，原因在于二头肌、三头肌的收缩舒展与手的打开握紧没有自然的联系。

2　思想方法

提高人工手的功能和控制效果应该从两个方面出发，一是需要多功能的假肢，当今已经有多自由度的人工手在研发中，例如 I-LIMB, Shadow Hand, Otto Bock Hand，截肢病人通过肌肉控制的方法实现一些运动功能的康复，存在较大可能性；二是需要新的控制策略。基于 EMG 模式识别的控制方法可用于控制多功能假肢。这种控制方法用双电极获取表面 EMG 信号，从而对手的伸开握紧、手腕的弯曲与伸展进行控制。病人经过训练后采集 EMG 信号，对信号进行分类并使用状态进行标记。提取出 EMG 特征，训练分类器进行运动类别的预测。该方法能够直接实现多自由度的控制，但受限于截肢患者不能提供足够多的 EMG 信号，例如，下前臂骨截肢患者较为幸运，其肢体仍存有部分前臂的肌肉,可用于恢复手和腕的运动。而对于肘部截肢患者,手臂仅有二头肌和三头肌,恢复手、腕和肘的运动存在较大困难。更为严重的情况是失去整个手臂的截肢患者,没有剩下的肌肉可用作信号源来恢复手、腕、肘和肩膀的运动。

对前臂骨截肢患者,使用 6 个电极从手臂采集肌电信号控制 6 种基本动作类别,可以得到很高的分类正确率(90%左右)。最近我们为前臂骨截肢患者开发了基于 EMG 模式识别的控制系统。在这个系统中,我们使用了 6 个电极在剩下的手臂上采集 EMG 信号,然后使用线性判别分析(LDA)方法对 EMG 信号进行解码,控制 3 自由度的机械手实现 6 个基本动作。但是对于在肘的上方截肢的患者来说,模式识别的控制方法受有限的剩余肌肉所限,因此我们需要其他的信息。

一个可行的解决方案是神经-机器接口技术。传统的神经-机器接口有 BCI(cortical neural signal)和 PNI(Peripheral neural signal)等。芝加哥康复学院提出了称为 Targeted Muscle Reinnervation(TMR)的新的神经-机器接口。

3 TMR

假设手臂截肢以后患者的剩余神经未受到损伤,残留的神经仍可传递来自大脑的神经信号,但神经信号仅能到达截肢末端。TMR 技术通过外科手术恢复截肢丧失的肌电信号。例如肩关节截肢患者,把胸肌分成四个不同的部分,把四个残余神经对应地转移到四个目标肌肉区域。经过 3~5 个月的时间,四种神经将会在四块目标肌肉区域重新生长。当患者想做一个动作的时候,大脑中产生的神经信号会往下传递到目标肌肉,刺激肌肉收缩。当肌肉收缩的时候,我们可以检测到很强的肌电信号。

在胸部的四个目标肌肉分别放四个双极电极采集 EMG 信号,用每个电极的 EMG 信号控制一种运动,截肢患者可以控制四种不同的手部运动。TMR 手术后,把 EMG 信号采集阵列贴在患者的胸大肌上,可以采集到足够多的肌电信号控制 16 种不同的动作。如果使用多于 100 个的双极电极控制 16 种运动可以获得很高的分类正确率。事实上,使用 12 个电极同样也可以达到很高的分类正确率,仅仅比使用 100 多个电极低了 3%。

4　应用

基于 TMR 接口和 EMG 模式识别算法，截肢患者可以控制人工手完成多种功能运动。如图 1 所示，左上角是捏一个小瓶盖，右上角是抓住并移动一个棍状物，底下是抓取一个纸盒子。

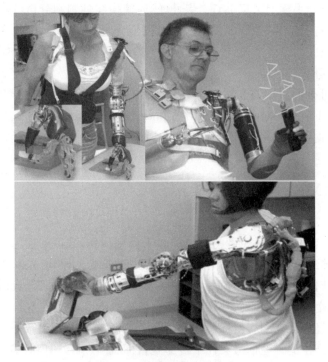

图 1　TMR 在实际中的应用

5　总结

人的肢体动作是由大脑、神经、肌肉活动协同完成的结果，当人的手或者部分手臂缺失，人会丧失很多功能，肌电控制的多功能假手被认为是比较好的一个解决方案。但是假手要比较好地模拟人手的动作将会变得有些复杂，甚至无法实现。多功能假手要完成人的很多动作就需要很多自由度，因此也需要比较多的控制源，然而截肢患者的控制源（肌肉）比较少，这也就对如何有效地控制造成了困难。

　　李光林教授主要研究了截肢患者如何控制多功能假手，通过 TMR 接口技术和肌电信号的识别，可以帮助截肢患者通过控制多功能假手完成一些动作，比如腕伸、腕曲、腕内旋、腕外旋、抓和捏。

　　TMR 技术克服了截肢患者肌电信号不足的问题，从而使人工假手的多功能控制成为可能。另一方面，TMR 手术操作可行性高，给众多的截肢患者重新获得灵巧的手带来了福音。

参考文献

［1］Farrel TR. Multifunctional Prosthesis Control: The Effect of Targeting Surface vs. Intramuscular Electrodes on Classification Accuracy, and Effect of Controller Delay on Prothesis Performance ［dissertation］. Evanston, IL: Northwestern University, 2007

［2］Farina D. Interpretation of the surface electromyogramin dynamic contractions. *Exerc Sport Sci Rev*, 2006, 34(3): 121 - 127

［3］Englehart K, Hudgins B, Chan AD. Continuous multifunction myoelectric control using pattern recognition. *Technol Disabil*, 2003, 15(2): 95 - 103

讲座人简介

　　　　　　Dr. Guanglin Li received the Ph. D. degree in biomedical engineering from Zhejiang University, China, in 1997. During 1999 - 2002, he worked as a Postdoctoral Research Associate in the Department of bioengineering, the University of Illinois at Chicago, on the studies of electrocardiography and electroencephalography inverse problems and cardiac electrophysiology. From 2002 to 2006, he was a senior Research Scientist at BioTechPlex Corporation, where he worked on the research and development of biomedical and biological products. From 2006 to 2009, he was with the Rehabilitation Institute of Chicago, where he was a Senior Research Scientist in the Neural Engineering Center for Artificial Limbs and with Northwestern University, where he was a Assistant Professor of Physical Medicine and Rehabilitation, Chicago. Since 2009, he has been with Shenzhen Institute of Advanced Technology (SI-AT), Chinese Academy of Sciences, where he is currently a Professor in the Research Center for Neural Engineering at Institute of Biomedical and Health Engineering. Dr. Li has published more than 50 peer-reviewed papers. Since 2006 he has been the Senior Member of the IEEE Society. He has served as a reviewer for more than 10 peer-reviewed international journals, as an associate editor for IEEE Transactions on Information and Technology in Biomedicine, and as a Member of the International Advisory Board of Physiological Measurement. His current research interests include neuro-prosthesis control, neural-machine interface, rehabilitation engineering, biomedical signal analysis, and computational biomedical engineering.

基于面部运动区和 Wernicke 区皮层微电极阵列信号的单词发音分类

Bradley Greger

(University of Utah, Department of Bioengineering, Salt Lake City,
UT 84112 - 9458)

摘要: 通过植入大脑皮层表面面部运动区和 Wernicke 区的微电极阵列记录到 ECoG 信号,我们可以高正确率地进行多单词的分类。研究表明,大脑皮层表面的空间分辨率能够达到毫米级别。此外,对于微电极的设计也进行了讨论,并根据数据分析结果提出了优化的电极设计方案。

关键词: 皮层脑电;语言解码;微电极

Using Micro-electrocorticography Recordings from Face-Motor Cortex and Wernicke's Area to Classify Spoken Words

Bradley Greger

(University of Utah, Department of Bioengineering, Salt Lake City,
UT 84112 - 9458)

Abstract: Electrocorticography (ECoG) grids record robust neural signals due to the close proximity of the electrodes to the cortical surface. However, standard clinical ECoG electrodes are large in both diameter and spacing so that a single electrode is integrating signals from approximately an entire cortical area. In micro-electrocorticography many micro-electrodes are closely spaced together in an area smaller than a single ECoG electrode. We are investigating the ability of micro-ECoG grids to record multiple independent neural signals from the cortical surface for use in neural prosthetic applications. This study describes the classification of spoken words using surface local field potentials (LFPs) recorded on subdural micro-ECoG grids. Data recorded from these micro-ECoG grids supported accurate and rapid classification of spoken words. Furthermore, electrodes spaced only millimeters apart demonstrated varying classification characteristics and strong correlations with different words, suggesting that cortical surface LFPs may encoded information with high temporal and spatial resolution. These results indicate the micro-ECoG is a promising technology for neural prosthetic applications, either on its own in cases where penetrating into the cortext is not desirable, or as an adjuct to penetrating micro-electrode arrays.

Keywords：Cortical eletroencephalogram, Linguistic decoding, Micro-eletrode

1　引言

通过大脑信号解码人类的语言、动作和思想，一直以来都是脑科学研究的重要目标。现阶段脑机接口技术的发展，已经初步实现了非人灵长类动物的动作解码，但是由于植入电极阵列使得被试需要承担一定的手术风险，植入式脑机接口的研究还停留在动物实验阶段。近年来提出的基于 ECoG 信号的半植入式脑机接口，提供了基于人类被试进行脑机理研究的可能性，已成为备受关注的研究课题。美国的 Leuthardt 等在 2004 年首次成功实现了基于 ECoG 脑机接口系统的在线一维鼠标控制，并认为 ECoG 信号具有连续实时控制的潜力，Leuthardt 等的研究开辟了 ECoG 脑机接口在线实验的先河，引发了 ECoG 脑机接口研究的热潮。

2　数据采集

目前参与 ECoG 的脑机接口研究的被试一般是因癫痫手术需要而植入了 ECoG 电极（图 1）的病人。由于癫痫手术本身的需要，病人一般会植入电极 5～7 天住院观察。目前对于 ECoG 脑机接口的研究主要集中在运动区，因此只有电极位置覆盖运动区的病人被邀请参与实验。实验需要向病人说明意图并征得病人同意，ECoG 电极的位置也完全遵照病人手术需要。考虑到术后病人的恢复，一般实验在电极植入后的 4～6 天进行。

图 1　ECoG 电极

我们的记录设备规格为 4mm×4mm，电极间距 1mm，是很微小的装置。在癫痫灶切除手术进行当中，病人需要植入 ECoG 电极阵列，来判断癫痫灶的具体位置，一般的医用电极直径在 3.5mm 左右，一个电极阵列有 64 个电极。我们的研究目的，一方面是探究大脑表面信息处理的尺度（scale），微电极的大小相比之下更接近于大脑皮层处理单元的大小，但我们想知道微电极要多么小，相距多么近才能记录到足够的信息；另一方面是我们能从这种微电极装置记录的 ECoG 信号中提取出哪些信息和特征，从而用来解码病人的语言和动作。

图 2 是微电极在电子显微镜下看到的样子，大概直径有 50μm。这也是从 19 世纪五六十年代制造的 ECoG 电极的基础上制作的，技术很简单但很实用。

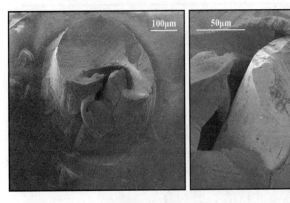

图 2　微电极细节放大

我们对这种装置进行了特性测试，图 3 是它的电阻特性谱，可见在 100Hz 的时候电阻在 297±283kΩ，还是不错的，可以很好地记录 ECoG 信号。此外，在我们需要的频段上的相位响应相对平稳。

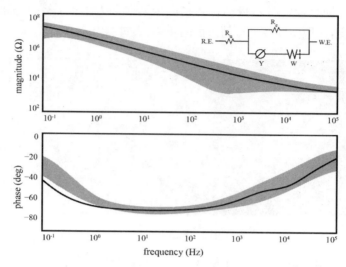

图3　电极电阻测试

图4是我们的一个被试,他植入了ECoG阵列电极,从电极引出的电缆线连接到Blackrock的数据记录装置进行数据的记录,被试需要根据一定的实验范式完成一些任务。这个被试正在进行伸抓实验,我们使用压力感应触摸板来记录手部的位置,我们也捕捉手指的运动,以及让他们说一些单词并用麦克风记录下来,传入Blackrock系统中,这样就使得语音信号和ECoG信号在时间上精确同步。令我们欣喜的是,实际上仅仅使用相对原始的信号,也就是在皮层表面记录下的局部场电位(LFP)信号,就可以对手部运动进行解码。

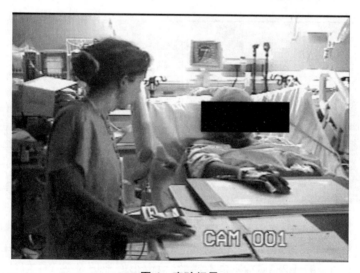

图4　实验场景

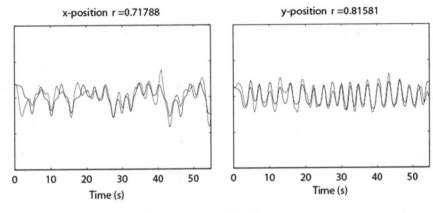

图 5 手部运动轨迹拟合

图 5 中深色的是手部控制光标的路线,浅色的是用 ECoG 信号解码出来的光标位置,可见通过我们的装置记录下来的信号能够解码手部在空间中的位置。从对二维坐标 X 和 Y 的解码轨迹中看出相关系数很高,在 0.7 和 0.8 左右,这样的结果对于一个简单的装置来说已经很好了,当然还有改进的空间。

在我们看来令人激动的一点,也是这种装置方便而独特的一点就是,它不需要植入大脑,跟那些需要植入的电极相比更加安全。尤其是对于像 Wernicke区域这样跟人类语言理解密切相关的区域,一旦稍有损伤会对病人造成极其严重的影响。

图 6 是一个病人大脑的 MRI 图,圆点是医用 ECoG 电极阵列,方块是微电极,这个病人的 Wernicke 和面部运动区被分别放置了微电极,面部运动区控制说话时舌头和嘴唇等的运动。

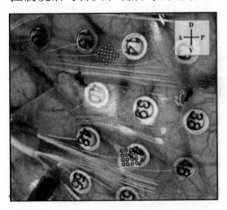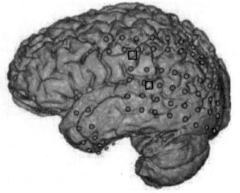

图 6 电极位置示意

3　实验设计

被试要做的就是在我们的语言指导下说话,图 7 是我们在 ICU 室里跟病人说话时的一段录音,先是一段正常对话,然后是任务开始,一遍一遍重复同样的单词,最后以我们的致谢结束。我们让病人重复这样十个独立的单词,分别是 yes, no, hot, cold, hungry, thirsty, hello, goodbye, more, less。

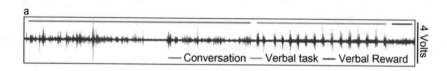

图 7　录音片段

图 8 是放置在脸部运动区的微电极的一个通道记录下的信号,可见在任务进行阶段,信号在时域上有明显的能量激发。一个令我们惊讶的现象是,放置在 Wernicke 区域的微电极单通道信号表明,Wernicke 区域的功率在任务进行时能量下降。各个通道信号的均值也显示出与上述一致的结果:在对话阶段 Wernicke 区域伴随着能量激发,但是一旦开始任务能量降低,直到我们致谢时被试自己组织语言才重新出现能量激发。

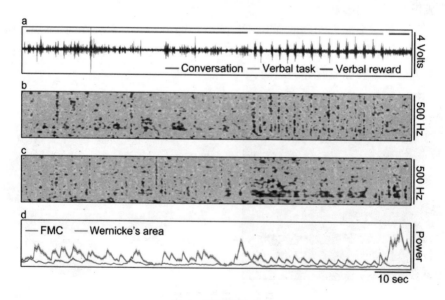

图 8　音频频谱分析

　　综观阵列中的每一个电极,图 9 中每个方块代表一个电极的频谱,左边是面部运动区的微电极阵列,右边是 Wernicke 区域的微电极阵列。由图 9 可见在很多频段上都有明显的特征,并且在时域上很活跃。图 9 是患者说"Thirsty"时的频谱图,如果你看患者说"Hello"时的频谱,就能发现其中的特征是不同的。因此,通过对频域、时域、空间、电极的能量变化的分析,就能发现一些很有趣的变化。

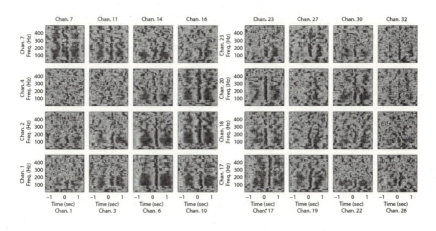

图 9　电极信号频谱

4　数据处理

　　信号的分析方面,我们试着研究一种算法(图 10),能够从这些神经信号中分析出被试要说什么。我们同时使用了时域和频域的信息,并且将时域上幅度的变化也加入进来,得到一个向量。对于每一个通道,我们都可以得到一个类似的向量。这是对于一个单词一次发音的特征。对于每一次发音都采取类似的方法,就能得到一个矩阵。对于这个矩阵,使用主成分分析(PCA)的方法使方差最大化,然后对每一次发音进行聚类,图 11 中红色和蓝色分别是Hungry 和 Thirsty 的聚类结果。对于训练数据,用相同的算法先是进行聚类,然后取每一类的中心;进行解码的时候使用这种算法,计算距离哪一类的中心更近,就判断出被试在说什么单词。

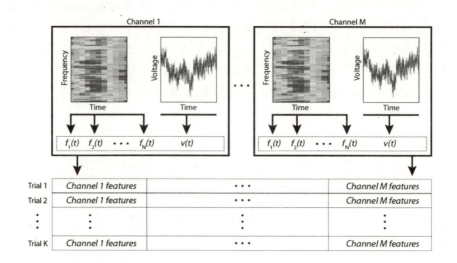

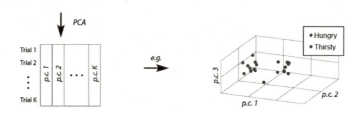

图 10　分类算法示意

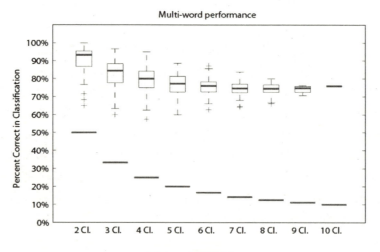

图 11　分类结果

5　结果分析

通过进行上述过程,我们成功地解码出十个单词,并且有的单词能够达到95%,总体能达到80%的正确率。我们对此很欣慰,认为这能够成为一种新的交流方式。

我们还对微电极阵列本身进行了研究,因为是一个新的研究方向,我们并不确定使用这种设备记录到的信息尺度(scale)是多大,如果我们知道了这一点,就可以设计更好的电极阵列。

图 12 是电极间距与相关系数的关系图,表示相隔不同距离的电极之间的相关性,在这里最大的距离是 4mm。对于广谱数据,相距 4mm 的电极之间依然很相关,但是对于高频 gamma 波段的数据,随着距离的增加,相关性递减,大致上是线性的趋势。

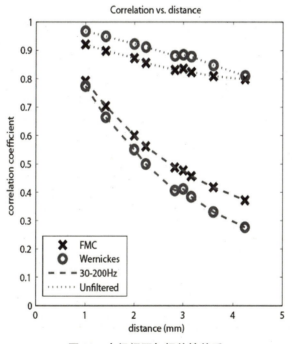

图 12　电极间距与相关性关系

同样,在空间尺度上,图 13 是每一个通道的分类正确率,由图可见尽管相距很近,每一个通道的分类正确率也有很大的差别,因此空间位置不同的电极记录着不同的信息,这也表明,皮层的信息表达单元是很小的。

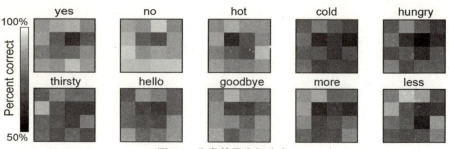

图 13　分类效果空间分布

我们又进一步观察了每一个通道的信号 PCA 后得到的第一主成分在时域上的表现。图 14 第一行是录下的音轨,中间的线段代表对应于每一个单词重复发音的数据段,可见对于某些数据段,PCA 的第一主成分有着明确并且鲁棒的特征。

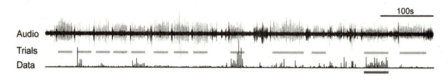

图 14　信号分析

我们将这段数据放大,同时也关注其他的电极,可见有些电极上,比如 1 号电极,跟某些单词有很强的相关性。就比如这段是被试重复一个单词的音频,1 号电极能够很好地反映发音情况,而 1mm 之外的其他电极上信息就少很多(图 15)。对于每个单词,都会有不同的空间模式。因此,我们使用这样的装置得到了很高的空间分辨率。

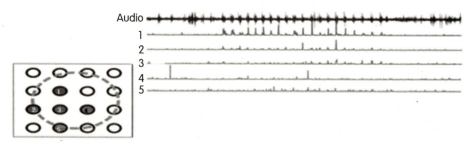

图 15　电极间信号对比

根据之前的实验,我们开始重新设计微电极,图 16 是我们现在使用的电极,其中有 16 个普通的 ECoG 电极,密密麻麻的连线每一条的终点都是一个微电极,共有 121 个微电极,整体是透明的。病人可以通过 ECoG 电极的信号探知癫痫灶的位置,而我们也可以通过微电极记录我们想要的信息。在病人的 MRI 图中可以看到微电极的位置在运动区附近，这是我们正在进行的实验。

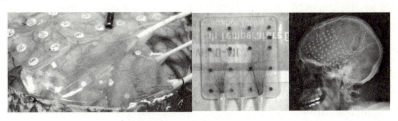

图 16　微电极设计

所以我们想做的，是通过最优电极的设计，采集到最多的信息；同时我们也不断改进算法，由于采集到的信号是非常不稳定的，每一次实验采集的信号都有变化，因此我们试图采用自适应的解码算法；同时也希望能够进行在线解码，通过我们的解码算法能够实时地给被试反馈。

总之，通过解码皮层表面的电位，我们能够成功地解码到目前为止 10 个单词，由于性能比较良好，所以我们会考虑解码更多的单词，比如 50 到 100 个，将会给一些病人带来方便。皮层信号在时域和频域的特征，既丰富也复杂，还有待进一步探索。我认为最重要的一个认识是，即使在皮层，信号的空间分辨率也能够达到毫米级。

讲座人简介

Research in the Greger lab focuses on understanding how information is encoded in neural structures and on how to apply this understanding to the treatment of human pathology. Electrophysiological recordings and electrical micro-stimulation are used to gain an understanding of how the nervous system processes information related to various sensory, motor, and cognitive functions. The results of these experiments are then used to guide implementation of neural prostheses for human patients. Neural prosthesis for the profoundly blind and amputees are currently being developed in the Greger lab through DOD and NIH funded projects. Electrophysiological research in humans aimed at understanding and treating epilepsy and other neural pathologies is also ongoing in the lab.

Dr. Greger has faculty appointments in Bioengineering and the Moran Eye Center at the University of Utah, and he collaborates closely with colleagues in the departments of Neurology, Neurosurgery, and Orthopedics. He earned his PhD from Washington University in St. Louis where he performed psychophysical experiments in both healthy subjects and patients with various motor pathologies, and performed electrophysiological experiments in behaving nonhuman primates. During his postdoc at Caltech he worked on decoding cognitive signals from the posterior parietal lobe for controlling neural prostheses. He consults for several medical devices companies and serves as the chair of Blackrock Microsystems scientific advisory board. He has 25 peer-reviewed publications, 1 issued patent and 4 pending patents related to neuroprosthetic applications.

mGRASP 技术"绘制"大脑环路

Jinhyun Kim

(Center for Functional Connectomics, Korea Institute of Science and Technology, 39 - 1 Hawolgok-dong, Seongbuk-gu, Seoul 136 - 791, Korea)

摘要：要理解神经元间的相互作用,必须研究神经网络中的突触连接。自 20 世纪以来,虽然神经科学家发明了各种先进技术对突触连接进行观察,但这些技术仍存在这样那样的缺点(如耗时、耗人力、准确度不高等)。为克服这些缺点,我们发明了一种观察小鼠大脑突触连接的新技术。该技术利用绿色荧光蛋白的分割组合,因此称为 mGRASP（mouse GFP-Reconstruction Across Synaptic Partners）。mGRASP 技术将 GFP 分割为两部分,一部分在前突触表达,另一部分在后突触表达,在突触形成的情况下,这两部分才结合,形成完整的 GFP,发出绿色荧光,从而可快速、准确地定位突触。我们成功地应用 mGRASP 技术观察到小鼠海马区内 CA3 到 CA1 的突触连接,由此,我们认为 mGRASP 将为了解大脑功能打开新的篇章。

关键词：mGRASP;神经元突触;大脑重构

Mapping Brain Circuitry with mGRASP Technology

Jinhyun Kim

(Center for Functional Connectomics, Korea Institute of Science and Technology, 39 - 1 Hawolgok-dong, Seongbuk-gu, Seoul 136 - 791, Korea)

Abstract：Investigating synaptic connectivity in neuronal circuits is necessary for understanding the functions of interacting populations of neurons. Over the past century, neuroscientists have put to use the most advanced techniques available to examine synaptic connectivity in neuronal circuits but newer, more powerful techniques are needed to overcome their limitations (i. e. labor and time intensity, ambiguity etc.). Here, we developed an alternative approach for mapping synaptic connectivity in the mouse brain. The GRASP (GFP-Reconstruction Across Synaptic Partners) system is based on the split-GFP reassembly technique that has been demonstrated in C. elegans. The non-fluorescent split-GFP fragments can reassemble into a functional fluorescent form only when presynaptic arbors and postsynaptic arbors are sufficiently close to permit synaptic membrane carrier proteins to span the intercellular gap (synapses). Thus, using this technique, fluorescence indicates, quickly, confidently, and with high spatial resolution, the locations of synapses. However, we reasoned that the GRASP technique will require modification before it can be used as a transmembrane proximity detector for visualizing synapses in the mouse brain, since synaptic architecture varies across organisms. Using in silico technology, chimeric presynaptic and postsynaptic mouse GRASP (mGRASP) components to accommodate the mouse synapse formed by thickened presynaptic and postsynaptic membranes with a synaptic cleft between them (average width of 20 nm) were designed. mGRASP was successfully applied to hippocampal CA3 - CA1, thalamocortical, and other circuitries in the mouse brain. We suggested that mGRASP will provide new avenues of insight into the brain functions by precisely characterizing synaptic connectivity of neuronal circuits in health and neurological disorders (e. g. Autism).

Keywords: mGRASP, Neurons synapses, Brain remodeling

1 引言

　　神经系统的网络重建是了解神经系统工作机制的基础。重建网络的方法主要有两种,一种是光学显微镜下的稀疏重建(Sparse Reconstruction),另一种是电子显微镜下的致密重建(Dense Reconstruction)。光学显微镜和电子显微镜各有优缺点。电子显微镜的分辨率很高,可以看到纳米级的结构,分辨每一个突触,但电子显微镜的设备和维护成本要远高于光学显微系统,而且样本准备和仪器操作繁琐,后期处理也极为复杂,且需要进行大数据量图像的配准和复杂的纹理识别等。比如线虫总共只有 302 个神经元,但用电子显微镜重建其整个神经网络花了 10 年左右。另外,电子显微镜下不能进行动态成像,无法观察神经系统在学习或病理条件下发生的变化。所以用电子显微镜重建更高级动物如小鼠这样复杂的大脑,只能局限于一个很小的区域。光学显微镜的重建成本要低很多,而且能观察更大的区域,但由于分辨率的限制观察不到突触。所以我们迫切需要一种标记小鼠神经元突触的新型分子生物学技术,可以用荧光分子观察神经细胞之间的连接进而建立大脑的神经传导网络。

2　mGRASP 技术

　　mGRASP 技术是一种基于绿色荧光蛋白的技术。绿色荧光蛋白(Green Fluorescent Protein,GFP)是一种在蓝色波长范围的光线激发下会发出绿色荧光的蛋白质。GFP 由 11 条 beta-strands 组成。mGRASP 技术将 GFP 分成 sp11 和 sp1－10 两部分,分离后的两部分各自都不发光。当两部分距离十分接近时,才会重新形成荧光蛋白并发出荧光。应用这种技术,可以在前突触表达一部分,后突触表达另外一部分蛋白,当两种分离的蛋白在突触膜上表达并且十分接近时,就可以用荧光信号检测出来(图 1)。

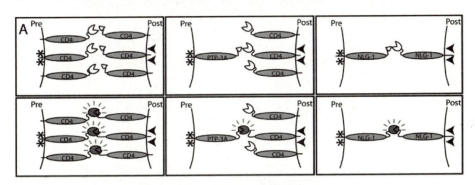

图 1　GRASP 工作原理图

　　我们对 mGRASP 技术在小鼠海马区进行了测试。海马是哺乳类动物中枢神经系统中脑的部分(大脑皮质)中进行过最为详细研究的一个部位,信息进入海马时由齿状回流入 CA3 到 CA1 再到脑下脚,在每个区域输入附加信息后在最后两个区域输出。利用 mGRASP 技术可以显示 CA3 区与 CA1 区的连接,分析 CA3 如何向 CA1 传导信号。为实现这个目的,我们可以在 CA3 中表达前突触 sGFP 部分,CA1 中表达后突触 sGFP 部分。完整的实验过程分为三步:GFP 分子分离;用子宫内电穿孔技术和病毒转染技术将两部分蛋白送到不同的区域;分析图像,重建神经结构,检测 sGFP 的荧光,分析分布情况。具体如下:

　　第一步,荧光分子经过设计后,其结合后长度刚好可填满突触间的150~200Å的空隙。前突触的膜用蓝色荧光标记,后突触用红色荧光标记,这样可同时得到前后突触的细胞结构。

　　第二步,怀孕15天的小鼠用子宫内电穿孔技术在CA1区注入后突触部分的DNA,出生后2个月,用病毒转染CA3区域,然后用Cre-LoxP重组技术可以实现基因的定时开关(图2)。

　　第三步,对小鼠大脑切片进行共聚焦扫描,得到海马区的图像(图3),图像中包括CA3区域的轴突结构、CA1区域的树突结构以及两者的连接突触。神经元图像拼接后把三个通道分开,得到轴突树突以及沿树突分布的突触。为了分析树突结构以及突触的分布,应用计算机程序对树突的三维结构进行数字化重建,将树突按照树的结构分为各种等级后(从根节点开始按分支层次),得到接收信号的突触在各种等级的树突上的分布。为了检测突触点,增大数字化重建后的树突半径并在增大的区域内检测突触,最后得到突触沿树突的整体分布。

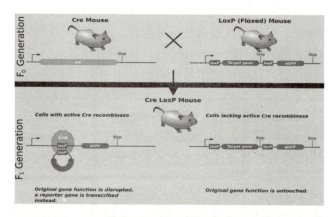

图2　Cre-LoxP重组技术

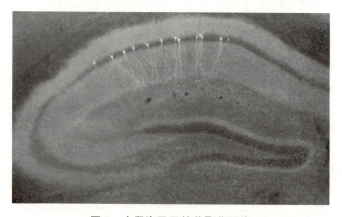

图3　小鼠海马区的共聚焦图像

3 总结

mGRASP 技术可帮助我们快速直观地重建出大脑的神经连接,了解神经信号传导的机制和动态过程,这必将成为神经科学中一项重要的工具。

参考文献

[1] Feinberg EH, VanHoven MK, Bendesky A, Wang G, Fetter RD, Shen K, Bargmann CI. GFP Reconstitution Across Synaptic Partners (GRASP) defines cell contacts and synapses in living nervous systems. *Neuron*, 2008, 57(3): 353 – 363

讲座人简介

For Master's thesis of Dr. Jinhyun (Jinny) Kim at the Sung Kyun Kwan University, Korea, she analyzed genomic sequences of oleosin in plant oil bodies and worked to improve nutrition by generating transgenic plants expressing recombinant oleosin with additional essential amino acids. After this experience with plant physiology, she moved to Germany to pursue what had become her strongest interest, Neuroscience. For her PhD thesis at the Max-Planck-Institute, she developed novel techniques to express a gene of interest in a desired region of rodent brain at a desired time, by using inducible transgenic mice (i. e. tTA and Cre-loxP systems) and virus-based expression systems (i. e. modified Sindbis and Adeno-associated virus). Bringing these techniques with her to the National Institutes of Health, USA, she investigated the subcellular distribution and dynamic molecular behavior of dendritic ion channels in hippocampus through imaging, molecular physiology, and other molecular biology tools. Most recently, at Janelia Farm Research Campus, HHMI, USA, she have begun to study complex subcellular and intercellular networks, focusing on basal patterns of synaptic connectivity and their modification by plastic mechanisms. Currently, she continues to develop new tools for brain circuitry mapping (mGRASP technology) at the Center for Functional Connectomic (CFC), KIST, Korea.

灵巧假肢设备的神经控制

Soumyadipta Acharya

(Graduate Program Director, Center for Bioengineering Innovation and
Design Assistant Research Professor in Biomedical Engineering
Johns Hopkins University Baltimore, MD 21218)

摘要:脑机接口(BMI)的最新进展已使得我们可以对机器假肢装置进行直接的神经控制。但是,对于是否可以通过对皮层信号的解码实现实时重建单个手指和手腕的灵巧运动目前还不是十分清楚。在该研究中,我们记录了当受训猕猴右手手指和手腕单独运动时,主运动皮层(M1)与任务相关的 115 个单个神经元的活动,并通过随机选择部分连续神经元,创建虚拟多单元组合,即体元,然后设计了使用人工神经网络的非线性分层过滤器,可实时对多个虚拟单元进行神经活动的异步解码,将解码结果用于驱动机械手的单个手指运动。

关键词:脑机接口;解码

Neural Control of Dexterous Prosthetic Devices

Soumyadipta Acharya

(Graduate Program Director, Center for Bioengineering Innovation and
Design Assistant Research Professor in Biomedical Engineering
Johns Hopkins University Baltimore, MD 21218)

Abstract: Recent advances in Brain-Computer Interfaces (BMI) have enabled direct neural control of robotic and prosthetic devices. However, it remains unknown whether cortical signals can be decoded in real-time to replicate dexterous movements of individual fingers and the wrist. In this study, single unit activity from 115 task-related neurons in the primary motor cortex (M1) of a trained rhesus monkey were recorded, as it performed individuated movements of the fingers and wrist of the right hand. Virtual multi-unit ensembles, or voxels, were created by randomly selecting contiguous subpopulations of these neurons. Non-linear hierarchical filters using Artificial Neural Networks (ANNs) were designed to asynchronously decode the activity from multiple virtual ensembles, in real-time. The decoded output was then used to actuate individual fingers of a robotic hand.

Keywords: Brain-machine interface; Decoding

1　引言

脑机接口(BMI)可以实现大脑和计算机或机械装置等外部设备之间的通信。对获取大脑皮层控制信号的研究主要集中于对背外侧运动前皮层(PMd)、主运动皮层(M1)和后顶叶皮层区(PPC)等区域神经元的整体解码。近几年来,通过对神经元进行解码来实现控制假肢的技术取得了巨大的进展,国际上有一些实验室已成功研制出非常灵巧的假肢,如约翰霍普金斯大学物理实验室的假肢,不仅从外观感觉上与人手十分相像,还可实现单个手指的控制和灵活运动。现在的挑战是:能否通过脑机接口实现一维控制、二维控制甚至是三维控制,使得该领域产生一个大的飞跃。另一方面,如何为大脑提供精细反馈也是一大挑战。

在讨论如何进行解码以及发展解码技术之前,需要考虑几个基本问题:初级运动神经元如何对灵巧手的运动信息进行编码? 能否根据运动神经元的活动建立模型解码出手指、手腕的灵巧运动? 能否建立模型对连续运动进行解码而不仅是单一的状态? 能否解码所有连续的手指、手腕的抓取运动而不仅是简单的上下左右运动?

2　神经元编码模型

神经元信号的行为是来自多方面的,有些是外部因素,如外部刺激,有些则是广泛的活动,如手指运动,当然还包括了其他因素,如其他神经元对整体的影响作用,以及神经元自己的历史活动。因此,可以从简单的问题开始,仅仅考虑单个神经元的行为,而忽略其他所有的因素,尝试着看是不是一个很好的模型。从统计考虑,电极中的大部分运动神经元信号符合一些概率分布。通过计算神经元放电频率,以各种影响因素为条件并给定时间 t,可以导出一个能用于每个神经元的通用线性模型:

$$\lambda(t \mid H_t) \stackrel{\Delta}{=} \lim_{\Delta \to 0} \frac{P(Y_{t+\Delta} - Y_t = 1 \mid H_t)}{\Delta} \tag{1}$$

实验中我们所使用的是一只经过多次训练的猴子,在实验中抓着一个操纵杆(图1),并弯曲伸展单个手指,当它看到 LED 面板上的灯亮后就会移动相应的手指。当给猴子的指令是弯曲它的食指时,可以发现和食指相连的手指也有弯曲,这和人手的情况是同样的,在弯曲中指时,食指和无名指也会有相应的弯曲。

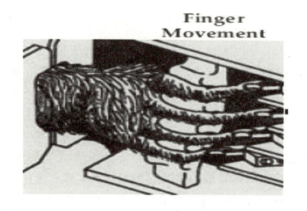

图1　猴子在实验中抓取的操纵杆

经过这些分析,可以提出一个基于实验范式的模型公式,是一个 M1 神经元放电频率的函数,x 对应于五个手指的位置。

$$\lambda(t) = f(x_1(t), x_2(t), \cdots, x_k(t)) \tag{2}$$

神经元放电频率符合泊松分布,可以导出一个广义拉格朗日(GLM)模型,该模型是一个基于泊松分布的通用线性模型,可用于建立单个神经元信号的行为:

$$\log \lambda_n(t) = \beta_0^n + \sum_{k=1}^{6} \beta_k^n x_k(t + \tau) \tag{3}$$

我们记录了两只猴子的神经元信号(一只 115 个,另一只 135 个),得到相关性的结果大约在 0.4 左右,相关性不高的原因是我们忽略了许多其他影响神经元行为的因素。

3 神经元解码模型

通过统计并观察神经元群体的活动,可以试着使用上述模型对手的运动进行解码,计算出每个手指具体的运动,但是结果并不理想。因此可尝试从一个简单线性模型转换到一个非线性模型,利用相同的实验系统(图2)和相同的神经元实时活动。

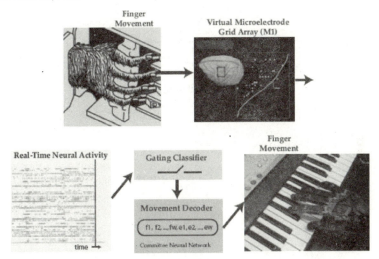

图 2　实验系统总览

考虑到神经元活动的非线性特征,这里采用了神经网络的非线性方法。该神经网络为一个两层网络,第一层用于检测运动意图,第二层用于检测运动类型,也即检测哪个手指在运动(图3)。

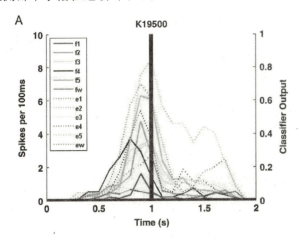

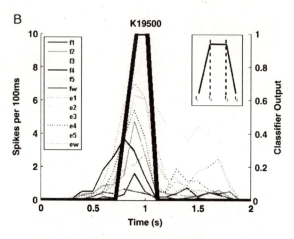

图 3　神经网络检测运动示意图和类型

我们记录了四只猴子的神经元活动,实验结果可以非常准确地预测出是哪个手指在运动(图 4)。

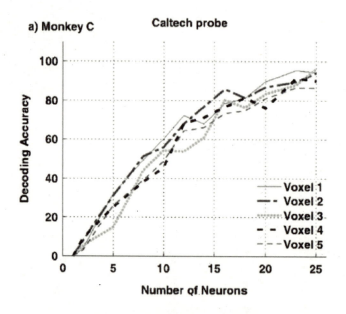

图 4　神经元解码的手指运动准确度

图 5 是使用了递归神经网络的非线性模型得到的结果,其中包括手指实际的运动以及解码得到的手指的运动,拟合结果非常好。从图中神经网络输出的波形也可以很明确地知道是哪个手指在运动。

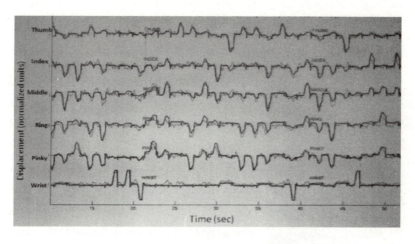

图 5　神经网络输出的手指运动波形

4　总结

该报告讲述了一些针对单个和群体神经元活动的简单线性算法，以及针对群体神经元活动的非线性算法，包括人工神经网络、递归神经网络模型。用这些算法在猴子身上进行的闭环实验，不仅包括手指和手腕的运动，也跟踪了猴子在伸手做复杂操作时整个手的运动。从实验结果得出的结论是，非线性模型的解码效果更好，更能够还原各个手指的实际运动，并实现对机械手运动的控制。

参考文献

[1] Acharya S, Tenore F, Aggarwal V, et al. Decoding Individuated Finger Movements Using Volume-Constrained Neuronal Ensembles in the M1 Hand Area. *IEEE Transactions on Neural Systems and Rehabilitation Engineering*, vol. 16, no. 1, february 2008

[2] Aggarwal V, Acharya S, Tenore F, et al. Asynchronous Decoding of Dexterous Finger Movements Using M1 Neurons. *IEEE Transactions on Neural Systems and Rehabilitation Engineering*, vol. 16, no. 1, february 2008

[3] Acharya S, Aggarwal V, Tenore F, et al. Towards a Brain-Computer Interface for Dexterous Control of a Multi-Fingered Prosthetic Hand. Proceedings of the 3rd International IEEE EMBS Conference on Neural

Engineering Kohala Coast, Hawaii, USA, May 2 - 5, 2007

讲座人简介

Soumyadipta Acharya (S'05) received the M. B. B. S. degree from Calcutta University, India, in 2000 and served as Resident House Physician in Cardiology at SSKM Hospital, Calcutta, India. He received the M. S. E. degree in biomedical engineering from the University of Akron, OH, in 2004. He is currently working toward the Ph. D. degree in biomedical engineering at the Johns Hopkins University, Baltimore, MD. His primary research interests are neural signal processing, brain-machine interfaces, and neuroprosthetics. Dr. Acharya has been the recipient of the Outstanding Graduate Student in BME Award ('04), University of Akron President's Letter of Commendation for Excellence in Research ('04) NASA Tech Briefs Award ('06), and the Outstanding Teaching Assistant Award ('07).

光基因技术静息位置细胞活动中的 PV 中间神经元

Sebastien Royer

(Center for functional Connectomics，WCI，KIST Seoul，Korea)

摘要：为了研究神经网络的实时动态特性，需要对神经元进行高时-空分辨率的刺激，同时对神经元集群的信号进行记录。多通道硅电极阵列能够对大量的单神经元进行检测。通过对含有光敏通道蛋白或者其他感受器的神经元进行刺激是研究局部神经环路的有效方法。这里我们介绍了硅电极阵列联合微光导管的光电极的制作方法，以及该光电极在小鼠和大鼠上的应用。此光电极的优点是能够在靠近记录位点的地方给予光刺激，从而增强了刺激的空间分辨率。

关键词：光基因；光敏通道蛋白 2；兴奋性；古细菌视紫红质；抑制性；光电极

Optogenetic Silencing of PV Interneurons During Place Cell Activity

Sebastien Royer

(Center for functional Connectomics，WCI，KIST Seoul，Korea)

Abstract：Recordings of large neuronal ensembles and neural stimulation of high spatial and temporal precision are important requisites for studying the real-time dynamics of neural networks. Multiple-shank silicon probes enable large-scale monitoring of individual neurons. Optical stimulation of genetically targeted neurons expressing light-sensitive channels or other fast (milliseconds) actuators offers the means for controlled perturbation of local circuits. Here we describe a method to equip the shanks of silicon probes with micron-scale light guides for allowing the simultaneous use of the two approaches. We then show illustrative examples of how these compact hybrid electrodes can be used in probing local circuits in behaving rats and mice. A key advantage of these devices is the enhanced spatial precision of stimulation that is achieved by delivering light close to the recording sites of the probe.

Keywords：Optogenetic，Light-sensitive channel protein 2，Ancient bacteriorhodopsin，Inhibition，Photoelectric pole

1　引言

当前神经科学研究面对的一项重要的挑战就是理清动物行为和神经元活动性之间的因果关系。在过去的十年，人们已经在这方面的研究取得了非常大的进步。研究神经元活动与特定行为之间的关系通常需要两个步骤：第一，确定引起特定行为的神经元的类型；第二，能操作这些细胞的动作电位发放。最近，分子光基因工具的发展为这方面的研究提供了有效的方法。通过在特定的神经元里面表达视紫红质通道蛋白2(channal rhodopsin 2，ChR2)，并用一定波长的光刺激这些神经元会使它们产生兴奋性的动作电位。而通过光刺激含有氯泵通道蛋白的神经元(halorhodopsins，NpHRs)可使它们的活动性迅速得到抑制。目前，很多人正利用这两项技术进行在体或者离体的研究，同时记录神经元的发放情况。通常记录位点和刺激位点的距离较远，因而需要较高功率的激发光才能满足刺激神经元的需要。这里我们介绍一种光电极的制作方法，以及该光电极在清醒动物上的应用：在光刺激的同时记录神经元集群的动作电位发放情况。

2　材料与方法

2.1　光电极的制作

为了获得能够同时进行光刺激和电记录的光电极，我们采用的方法是将硅电极与光纤导管耦合在一起(图 1)。我们采用 NeuroNexus 公司生产的硅电极探针作为记录电极。该探针通常有四个或者八个针脚，每个针脚上有八个记录位点，针脚之间的距离是 $250\mu m$。同时我们采用了 Thorlabs 公司生产

的单模光纤作为光刺激源(直径 125μm)。首先用氢氟酸腐蚀光纤 15～
30min,使光纤的直径达到 5～20μm。然后将光纤尖端平铺在硅探针的针脚
上,用环氧树脂将两者粘住,最后用紫外线照射使环氧树脂固定。光纤末端和
记录位点之间的距离是影响记录神经元效果的重要因素(图 2),通常间距在
100～300μm 之间。为了记录到刺激部位上面神经元的发放情况,光纤的尖
端要比硅电极针脚尖端短 100μm(图 2C)。

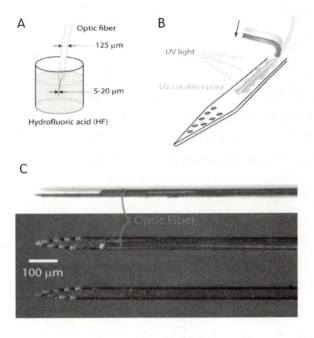

图 1　光电极的制作图

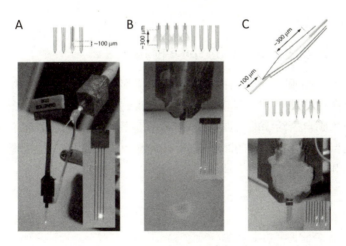

图 2　设计不同的光电极阵列图

2.2 大鼠实验

构建含有 ChR2-GFP（绿色荧光蛋白）的腺相关病毒（adenoassociated virus，AAV），并将此病毒载体注射到大鼠海马的 CA1 区。注射病毒 10 周后，训练大鼠在"8"字迷宫里面按照正确的路线行走。病毒注射 12 周后，利用立体定位仪将光电极植入到含有病毒的 CA1 锥体细胞层。用 473nm 的蓝光照射含有 ChR2 的神经元，同时记录神经元的电活动情况。

2.3 小鼠实验

构建含有 NpHR-GFP 的腺相关病毒，并将此病毒载体注射到含有 PV-Cre 转基因小鼠的背侧海马的 CA1 区。将小鼠的头部固定在立体定位装置上，训练清醒状态的小鼠在跑步机上行走。病毒注射 3～6 周后，利用立体定位仪将光电极植入到含有病毒的 CA1 椎体细胞层。用 561nm 的黄光照射含有 NpHR 的神经元，同时记录神经元的电活动情况。

2.4 数据处理与分析

神经信号通过 128 通道的 DigiLynx 系统获得，采样频率为 32.552kHz。原始信号在离线情况下，经过高通滤波（0.8～5kHz）检测到 spike 信号，经过低通滤波（0～500Hz）获得局部场电位信号（LFP）。利用 KlustaKwik 方法对 spike 信号进行自动分类，并利用 Matlab 对信号进行后续处理。

3 结果与分析

3.1 光激活 ChR2 时的 spike 检测

对大鼠 CA1 区进行短时的脉冲蓝色激光照射（5ms 波宽，1Hz）和长时的正弦蓝光照射（5Hz），可以检测到不同的 spike 发放情况（图 3 和图 4）。将这项技术应用到自由活动的大鼠身上，可以在迷宫中的特定部位激活相应的细胞。在大鼠经过迷宫角落的时候进行光刺激，可以看到有三个细胞被光激活（cell1，cell3，cell4），如图 5 所示。

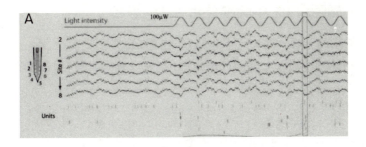

图 3　对 ChR2 神经元进行正弦蓝光照射

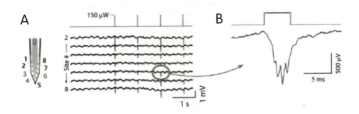

图 4　对 ChR2 神经元进行短时脉冲蓝光照射

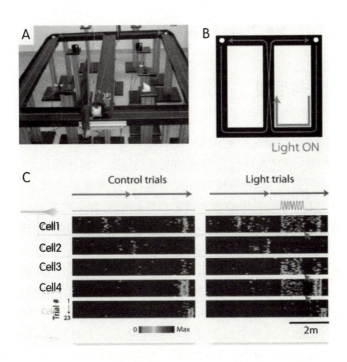

图 5　大鼠迷宫实验及位置特异性细胞刺激响应图

3.2　光照对 PV 中间神经元活动的静息

在 PV 中间神经元（parvalbumin，小白蛋白，属于钙结合蛋白超级家族）中表达 NpHR，用 561nm 的黄光进行照射，可以看到有些神经元的信号被静息了（图6）。

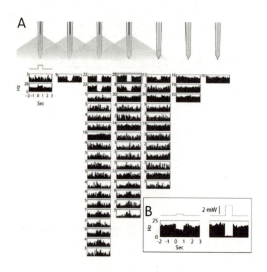

图6　NpHR 介导的 PV 神经元的激光沉默图

4　讨论与展望

光基因技术可以研究位置细胞（place cell）的生理特性。通过对大鼠位置细胞进行光刺激，研究大鼠位置细胞的电活动情况。相比大鼠，小鼠头部带有硅电极的记录装置太大，不易找到位置细胞，因此需要对小鼠进行头部固定，使小鼠在一个 1.5m 长的传动带上行走。传动带上有一些像刺点、粘扣、圆点等作为接触时信号提示点（cue）。当小鼠经过传送带上的提示点时，通过记录神经元发放情况，同样可以找到小鼠的位置细胞。

利用光基因技术还能对神经元的类型进行鉴定，例如很容易鉴别出小鼠的 PV 中间神经元。此外，利用该技术还可以研究整个网络的反馈情况，比如研究 PV 细胞对整个网络正常的活动影响，以及位置野（place field）发放率是如何改变的等。如果将整个位置野发放情况分为 5 个时间窗，分别计算每个时间窗的平均发放率（经过标准化处理），发现光照组的发放率均高于对照组的。此外，还可以对中间锥体神经元的 spike 发放时间进行研究。在局部场电位中存在 theta 节律,锥体神经元与 theta 节律存在一定的时间关系,但是这种时间关系不是相位固定的。在位置场细胞中,很容易观测到 theta 振荡(theta cycle)中出现的 spike,而位置野末端细胞的 spike 的相位提前了(theta phase precession)。其他的神经元也存在类似的现象,这个现象产生的机制目前尚不清楚。

参考文献

[1] Royer S, Zemelman BV, Barbic M, *et al*. Multi-array silicon probes with integrated optical fibers: Light-assisted perturbation and recording of local neural circuits in the behaving animal. *European Journal of Neuroscience*, 2010, 31: 2279 - 2291

[2] Royer S, Sirota A, Patel J, *et al*. Distinct representations and theta dynamics in dorsal and ventral hippocampus. *The Journal of Neuroscience*, 2010, 30: 1777 - 1787

讲座人简介

Dr. Sebastien Royer received his master degree in 2001 from the University Pierre et Marie Curie/ENSPM. He pursued his academic formation at Université Laval, Canada, where he obtained a PhD in Chemical Engineering in 2004. After one year postdoctoral stay at the French Petroleum Institute, he got a position of Associated Professor at LACCO, University of Poitiers, France. His current research is focusing on the computation of hippocampal circuits, large scale single units and EEG recordings in behaving animals, in addition of optogenetic.

面向神经假体和神经机器人的大脑微刺激

徐韶华

(Department of Physiology and Pharmacology, The State University of New York Downstate Medical Center, 450 Clarkson Ave, Brooklyn NY 11203)

摘要：人和动物都依赖感觉反馈对外部环境作出响应。神经功能失调如脊髓损伤会破坏大脑与躯干之间的联系,常导致患者丧失感觉运动功能且无法恢复。脑机接口(Brain Machine Interface, BMI)领域的最新研究进展表明,通过将大脑活动转化成运动命令,用于驱动人工设备,可修复基本的运动功能。在构建闭环 BMI 系统中,体感反馈是必需的环节。我们致力于在 BMI 系统中充分利用体感反馈,以大鼠和猴子为实验对象,试图实现利用中枢体感通路上的电刺激替代自然状态下的体感反馈。在主体感皮层(S1)的前掌对应区,使用多电极分别记录自然触摸和对 VPL 或 S1 的微电刺激条件下的神经集群反应信号,实验结果显示参数优化的微电刺激可产生与自然触摸类似的皮层神经反应。通过行为辨别任务实验,我们还研究了利用大脑微刺激给予大鼠脑部提示的可能性。此外,我们研发了"大鼠机器人",通过在特定脑区中使用微电刺激替代条件反射式的提示和奖赏,实现导航。综上所述,实现体感假体具备可行性,但还需要经精神物理学的进一步证实。

关键词：体感反馈;神经假体;大鼠机器人;大脑微刺激

Brain Microstimulation for Neuroprosthesis and Neuro-robotics

Shaohua Xu

(Department of Physiology and Pharmacology, The State University of New York Downstate Medical Center, 450 Clarkson Ave, Brooklyn NY 11203)

Abstract：Humans and animals rely on sensory feedback to act on their environment. Neurological disorders such as spinal cord injury can disrupt the link between brain and peripheral limbs, leading to usually unrecoverable loss of sensorimotor functions. Recent advances in brain-machine interface (BMI) have shown it is possible to restore basic motor functions by directly translating brain activity into motor commands enacted by artificial actuators. However, somatosensory feedback has not yet been fully functionally incorporated into the BMI even though it would be essential for optimal BMI control. Here I like to present our work on the feasibility of substituting somatosensory feedback through electrical stimulation in the central somatosensory pathway in rats and monkeys. Multi-site recording in the primary somatosensory cortex (S1) was used to study neuronal ensemble responses to discrete natural touch of the forepaw and electrical stimulation in the VPL or S1. A statistical comparison showed that optimized stimulation parameters yielded cortical neural responses that were remarkably similar to those evoked by natural touch. We also investigated how well a rat can utilize percepts of brain microstimulation as cues in behavioral discrimination tasks. We report the development of a "robo-rat", in which mild microstimulation through electrodes in appropriate brain locations substitutes for both conditioning cues and rewards. Rats were found able to use the percepts of brain microstimulation effectively and efficiently in a navigation task through a wide range of real world terrain. Taken together, our work suggested the feasibility of somatosensory prosthesis, which needs further confirmed by psychophysical studies.

Keywords：Somatosensory feedback, Neuroprosthesis, Robo-rat, Brain microstimulation

1 引言

脑机接口为大脑功能修复带来了希望。大多数研究集中于如何解码大脑的运动意图，进而控制机器手臂。我们的工作旨在如何将体感假体引进到大脑与肢体之间的神经通路中。日常生活中，体感反馈信息是很重要的，缺失了体感反馈的瘫痪病人将会无法自理。脊椎损伤、脑卒中或其他神经疾病都可能会导致体感功能的丧失。目前已有一些研究结果证明了研制体感假体的可行性，比如人工耳蜗、体感通路上的大脑微刺激等。

2　体感反馈

为此，我们以大鼠和猴子为模式动物进行了研究。腹侧后外侧（ventral posterolateral，VPL）丘脑为大脑皮层提供了躯干和肢体的重要体感信息，我们对大鼠的 VPL 进行了体感剖析，并在猴子的 VPL 区域和体感皮层植入犹他电极，进行了一系列实验，包括对 VPL 区域和体感皮层施加微电刺激后，记录体感皮层的反应信号。通过比较自然刺激与电刺激下的神经细胞反应，发现两种结果非常接近。刺激大鼠第二根手指并记录神经元发放的时间序列，比较机械触摸与电刺激 VPL 后的神经元发放信号，发现两者十分相似。我们还比较了刺激 VPL 区域和刺激运动皮层的神经元反应，结果表明对 VPL 区域的电刺激得到的效果更佳。而对运动皮层的刺激则显示出抑制性，即电刺激后的运动皮层神经元的响应比未受刺激时减小。采用因子分析（factor analysis）方法处理后的结果显示，自然触摸和电刺激所引发的神经集群响应基本落在了因子空间的相似区域。

3　大鼠机器人

我们利用电刺激技术实现了遥控大鼠导航的系统（图 1）。该系统可通过

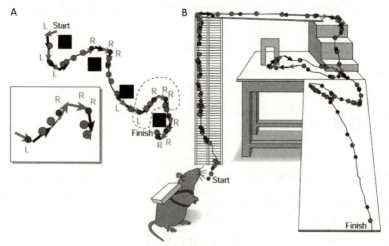

图 1　大鼠导航示意图

一台笔记本电脑向大鼠背负的控制背包发送信号,控制大鼠的运动,完成复杂环境的导航。植入到大鼠大脑中的电极分为三个,一个植入到 MFB 区域,使得大鼠在强化训练中获得"奖赏",另两个植入到接受左右胡须感觉输入的体感皮层中,使大鼠获得转向的提示。经过训练后的大鼠能够完成左右转向、前行等行为。导航实验中,我们设置了障碍和路线后,训练大鼠完成既定任务。未经训练的大鼠不能完成导航任务,而训练后的大鼠则可按照矩形路线行走或攀爬楼梯。这个成果引起了广泛关注,《纽约时报》、《探索》等杂志均有报道。

4 总结

经过优化的腹侧后外侧丘脑(VPL)刺激可以产生与自然刺激相类似的反应,为研制体感假体提供了可行性。用类似的技术可实现"大鼠机器人",即借助脑部电刺激,实现复杂环境中大鼠的运动导航。

在接下来的工作中,我们可进一步比较外周神经系统和中枢神经系统的精神物理学效应,推动体感假体的发展。

参考文献

[1] Francis JT, Xu S, Chapin JK. Proprioceptive and cutaneous representations in the rat ventral posterolateral thalamus. *J Neurophysiool*, 2008, 99(5): 2291 − 2304 [co-first author]

[2] Hermer-Vazquez LL, Hermer-Vazquez RW, Rybinnik I, Greebel G, Keller R, Xu S, Chapin JK. Rapid learning and flexible memory in "habit" tasks in rats trained with brain stimulation reward. *Physiol Behav*, 2005, 84(5): 753 − 759

[3] Xu S, Talwar SK, Hawley ES, Li L, and Chapin JK. A multi-channel telemetry system for brain microstimulation in freely roaming animals. *J Neurosci Methods*, 2004, 133(1 − 2): 57 − 63 (Published online December 2, 2003)

[4] Talwar SK, Xu S, Hawley ES, Weiss SA, Moxon KA, and Chapin JK. Rat navigation guided by remote control. *Nature*, 2002, 417: 37 − 38

讲座人简介

Dr. Shaohua Xu is a research scientist and instructor at State University of New York Downstate Medical Center, Brooklyn, NY. He received his bachelor degree and master degree in Shanghai Medical University and his Ph. D. degree in State University of New York Downstate Medical Center. He is a member of the Society of Neuroscience. He has a patent of Method and Apparatus for Guiding Movement of a Freely Roaming Animal Through Brain Stimulation. He has published papers on Nature, Journal of Neuroscience, Physiology & Behavior and Journal of Neurophysiology.

基于近红外功能成像技术的脑机接口

Hasan Ayaz

(School of Biomedical Engineering, Science & Health Systems,
Drexel University)

摘要:近红外(fNIR)功能成像系统广泛应用于非植入式脑功能研究。fNIR 利用氧合血红蛋白和脱氧血红蛋白的不同光学特性来检测两者的浓度变化,从而间接监测大脑活动。Drexel 大学的研究团队开发了基于 fNIR 的自调节型脑机接口(BCI)系统,用以实现人脑与计算机的交互活动。该系统属于非植入式 BCI 系统,具有便携、安全、低廉和无创等优点。本文从光学脑成像的背景着手,介绍了 fNIR 技术在脑机接口及认知神经科学领域的独特应用。

关键词:功能性近红外光谱(fNIR);光学脑成像;脑机接口(BCI);认知神经科学

Functional Near Infrared Spectroscopy based Brain Computer Interface

Hasan Ayaz

(School of Biomedical Engineering, Science & Health Systems,
Drexel University)

Abstract: Functional Near Infrared (fNIR) spectroscopy based optical imaging systems have been widely used in functional brain studies as a non-invasive tool to study changes in the concentration of oxygenated hemoglobin (oxy-Hb) and deoxygenated hemoglobin (deoxy-Hb) due to their different optical properties. Based on the fNIR technique, the team from Drexel University has developed a self-regulated, fNIR-based Brain Computer Interfacing (BCI) system to achieve the interaction between the human brain and the computer. The fNIR is portable, safe, affordable and negligibly intrusive since it belongs to the noninvasive BCI system. This paper begins with the background of the optical brain imaging, and thus interprets the specialty of fNIR's application in the field of BCI and cognitive neural science.

Keywords: fNIR, Optical brain imaging, BCI, Cognitive neural science

1　BCI 及 fNIR 技术的研究意义

脑机接口(Brain-Computer Interface, BCI)一方面是为瘫痪、残障人士提供一种辅助交流及方便生活的手段,包括面向感官障碍者的交流手段,以及面向运动功能障碍者的互动途径。另一方面,BCI 还有更多的应用领域,比如可以从自调节式脑活动的角度为一些脑部疾病的治疗方案提供生物反馈信息,这些疾病包括注意力缺陷性多动症(Attention-Deficit Hyperactivity Disorder, ADHD)、焦虑症、抑郁症以及其他精神病症。此外,BCI 系统可以作为一种额外的互动方式,提高人的认知能力与水平,表现为两个方面:一是在临床上应用于中风等神经系统疾病患者的康复工程中,二是为健康人群提供游戏及多媒体娱乐方面的一种新途径。

BCI 系统可分为植入式与非植入式两大类。其中非植入式 BCI 具有安全、方便等优点,避免了电极植入机体及相应手术带来的安全风险,因而目前应用于人体实验的 BCI 系统基本上都是非植入式的。然而,非植入式 BCI 目前的发展水平并不能够完全满足人们的需求。其中非植入式 BCI 最大的障碍是信号传感模块的问题,缺乏一种安全、准确、且系统鲁棒性良好,同时相对环保的大脑信号传感方式。多模态采集传感平台很可能是未来解决这一问题的有效方式;不过前提是在此之前我们研究设计出各种基于不同原理的信号采集与成像方法,并且它们在 BCI 领域的潜在应用空间已被充分发掘,这些还需要多方面的努力。

功能性近红外谱(fNIR)技术就属于这样一种大脑信号传感模式。fNIR 技术是一种新型、便携式、易获得的成像技术,它能够以非植入式方法提供大脑活动的客观反映,是一个年轻而充满潜力的发展方向。本文所探讨的便是 fNIR 大脑监测技术在新型 BCI 设备下的实现,并且构建带有反馈的闭环系统,使得用户与计算机之间的交流与互动得以增强实现(如图 1 所示)。

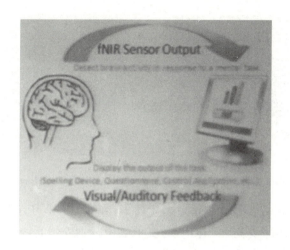

图 1　一种 fNIR-BCI 实现途径

2　光学脑成像与血动力学相关技术

以 fNIR 为代表的光学脑成像技术能够监测脑组织中血动力学参数的变化,是一类新兴的技术。fNIR 相对头皮脑电(EEG)具有更高的空间分辨率,相对 PET、SPECT、fMRI 等技术具有更高的时间分辨率。作为一种非植入式方法,fNIR 技术具有安全、经济、便携等优点,其系统模块可做成手机大小,十分方便;并且 fNIR 技术易与其他模态的传感检测技术相结合。

关于 fNIR 技术的成像原理,从物理学角度,可以由改进的 Beer-Lambert 定律来解释。一束自然光照射一块充满流体的人体组织时,只有一部分具有特定波长的光能够穿透人体组织并达到另一面。例如,将一只四指并拢的手伸进日光灯的光源处,如图 2 所示,只有红色系的光能够穿透手指并且在背向光源的一面可见。这是因为其他色系的光要比红色系光波长更短,均可被人体血液中的血红蛋白吸收;而红色系光能够穿过生物组织并在另一边显现。现已探明这一段能够穿透人体组织的光波所处的波长范围为 $700 \sim 900 \text{nm}$。在此我们主要考察氧合血红蛋白(HbO_2)与解氧合血红蛋白(Hb)对上述波段光束的作用;因此,可利用光束对不同人体组织的特定穿透作用,根据改进的 Beer-Lambert 定律,来测算该组织中血液内 Oxy-Hb 与 Deoxy-Hb 的浓度变化。

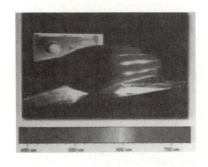
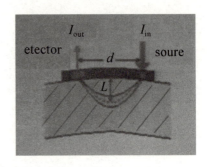

图 2　改进的 Beer-Lambert 定律的实例　　图 3　改进的 Beer-Lambert 定律模型图

改进的 Beer-Lambert 定律可表述如下：考虑如图 3 所示的简单模型，其中中间大块阴影区域表示人体组织，其上表面分别安置光源和光接收器。光源发射一束单色光到组织中，会被具有一定密度的人体组织所吸收；但人体组织并不是密度均匀的，因此光线在组织中传播时会发生折射。其中一部分光经过多次折射后最终返回上表面，被光接收器检测到，其光强对于不同波长的光来说通常是不同的。定义"吸光度（Optical Density，OD）"概念，即对于波长为 λ 的光波，该组织的吸光度为

$$OD_\lambda = \lg\left(\frac{I_{in}}{I_{out}}\right) \approx \varepsilon_\lambda \cdot c \cdot d \cdot DPF + G$$

其中入射光强与光接收器所检测到的光强分别用 I_{in}、I_{out} 表示，c 为组织中液体的浓度，d 为光源与光接收器之间的距离，ε_λ 为组织的吸光系数；DPF 为差分路径因子，取决于入射光波长以及组织的光学特性；G 为某固定常数。

实际研究中，人体组织内的血液是时刻流动的，因而不同时刻组织的液体浓度以及吸光度等参数都有微小的变化，这些变化的主要因素便是血液中 HbO_2 与 Hb 的浓度变化。保持入射光强 I_{in} 不变，对于不同的两个时间点，基准时间点 rest 与测试时间点 test，考察组织对波长为 λ 的光的吸光度改变：

$$\Delta OD_\lambda = \lg\left(\frac{I_{rest}}{I_{test}}\right) \approx \varepsilon_\lambda^{Hb} \cdot \Delta c^{Hb} \cdot d \cdot DPF + \varepsilon_\lambda^{HbO_2} \cdot \Delta c^{HbO_2} \cdot d \cdot DPF$$

其中，吸光度 OD 可通过取对数计算得出；距离 d 很易获得；吸光系数以及 DPF 对于特定组织而言也可测出；事实上，上式中只有 Δc^{Hb} 与 Δc^{HbO_2} 两个待求的未知参数。这样，我们可以选取两组不同的光波波长 λ_1、λ_2，在 rest 与 test 两个时间点上，两组光束同步射出，然后通过联立方程组求解 Δc^{Hb} 与 Δc^{HbO_2}，即得出两个时刻之间 HbO_2 与 Hb 的浓度变化情况，从而实现利用改进的 Beer-Lambert 定律和 fNIR 技术进行血动力学脑功能成像。

3   fNIR 信号特点及相关研究工作

由于 fNIR 所依托的基本原理及所利用技术手段的特点，fNIR 信号从时间和空间两个维度上都能得到较好的展开。图 4 显示了 fNIR 信号在时域上展开的波形图，图中的曲线是光接收器所得光强随时间变化的曲线。曲线中较大的尖峰反映被试的身体运动，曲线中包含着频率较高的、有规律的波动是心动周期的反映；需要指出的是，此时心动波形并不是我们想要的信息，因而该波形需要与设备噪声、呼吸干扰等信号一起消除或减弱。

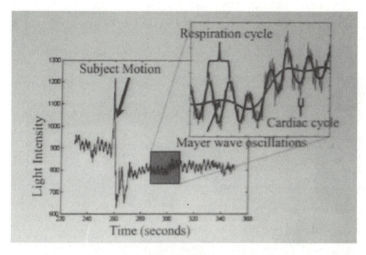

图 4   fNIR 时域信号波形图

当前，国际上有许多研究团队研究基于 fNIR 的 BCI 技术，主要分为两大方向：一是肢体运动及运动想象方面的研究与实现，目前有 Sitaram 团队、Abdelnour团队等，主要研究与 EEG-BCI 相类似的非植入式 BCI 范式；二是搭建 fNIR-BCI 系统，实现各种认知任务的执行，诸如心脑算术、口语流利表达等，目前主要有 Ayaz、Naito、Bauernfeid 等团队在进行此方向的研究。

在 BCI 领域研究中频繁和 fNIR 结合的技术是 fMRI。国际上针对 fMRI 的研究主要包括两个方面，一是肢体运动及运动想象方面的研究与实现，目前主要有 Yoo、Sitaram 与 Weiskopf 等团队从事此类研究；二是在局部脑活动实验中为自调节式被试训练系统提供神经反馈的 fMRI 技术，目前主要有 Fitzgerald、Weiskopf、Christoff 等团队进行此类研究。

4　自调节式 fNIR 系统

上一节提及有研究团队利用 fMRI 技术提供反馈以实现自调节式的 BCI 训练系统。但是，与其他非植入式技术相比，fMRI 设备相对庞大，代价较高昂。因此，Ayaz 团队尝试采用轻便而经济的 fNIR 技术代替 fMRI 来完成自调节式 BCI 系统的任务。该团队进行了一些设计实验，验证了 fNIR 技术在此项系统工作中的可行性，并且实现了简单的基于 fNIR 的自调节式 BCI 训练系统。实验范式可概括如下：实验被试戴上 fNIR 相关传感设备，在实验中注视电脑屏幕，整个过程经历"静息-任务-静息"三个状态，其中任务是令被试凝视屏幕中央出现的纵向亮条管，并努力用意念使管中的蓝色填充物向上充满，如图 5 所示。实验证明了 fNIR 在自调节式 BCI 训练系统中的可行性。

图 5　fNIR 系统实验实景图

5　总结

不同于 EEG、ECoG 等捕捉大脑电活动信号的技术以及 fMRI 等磁共振成像的技术，功能性近红外（fNIR）光谱技术利用红外光学成像，考察对象主要是大脑皮层的血流动力学特征，fNIR 有助于揭示大脑皮层的功能性激活与认知的关系及其机制。作为一种非植入式技术，它具有安全无损的优点，同时也具有设备轻便、成本较低等优势，在 BCI 领域的研究中得到了一定的施展空间。当然，与其他非植入式 BCI 一样，fNIR 技术所得到的信号分辨率、脑部信号定位的精确程度与植入式 BCI 相比仍有一条很难逾越的鸿沟。然而我们也要看到,随着光学传感技术的提高以及生物信号分析处理能力的逐步增强,基于 fNIR 技术的 BCI 本身的独特优势逐渐显现。而且,从应用层面上讲,目前 fNIR 已有多方面的应用和商业化产品,这也能够很好地体现这一点。

参考文献

[1] Huppert TJ, Diamond SG, *et al*. HomER: a review of time-series analysis methods for near-infrared spectroscopy of the brain. *Appl Opt*, 2009, 48(10): D280 - 298

[2] Ayaz H, Sheworkis P, *et al*. Assessment of cognitive neural correlates for a functional near-infrared based Brain Computer Interface system. Foundations of Augmented Cognition. *Neuroergonomics and Operational Neuroscience*, 2009: 699 - 708

[3] Ayaz H, Izzetoglu M, *et al*. Detecting cognitive activity related hemodynamic signal for brain comupter interface using functional near infrared spectroscopy. 3rd IEEE/EMBS Conference on Neural Engineering, 2007: 342 - 345

讲座人简介

Hasan Ayaz, PhD is a Biomedical Engineer. He received his BSc. in Electrical and Electronics Engineering at Bogazici University with high honors and PhD degree from Drexel University where he developed enabling software for functional Near Infrared Spectroscopy (fNIR) based brain monitoring instruments. This technology has been licensed by fNIR Devices LLC and is being distributed to research lab worldwide by Biopac Systems, Inc. As an extension to this, he worked on a portable-handheld medical device (InfraScanner) that utilizes fNIR to detect hematoma in head trauma patients. InfraScanner received EID (Excellence in Design) 2007 Gold Award and Frost & Sullivan North American Product Innovation of the Year Award in 2008. InfraScanner is currently pending FDA approval but has been deployed overseas and is already saving lives in Africa and Europe.

神经集群活动的低维表征

Sung Phil Kim

(Department of Brain and Cognitive Engineering Korea University,
Anam 5ga，SeongbukGu，Seoul，Korea)

摘要：长期植入式多通道微电极阵列正成为记录灵长类动物皮层脑电活动的常规方法，该技术能记录上百个神经元的胞外电信号。神经活动往往比刺激表现出更多的动态特性。多通道神经元集群的动态变化携带了大量的信息。传统的神经信号分析方法不能适用于单次实验神经元集群动态活动的分析。本文介绍了一种将神经元集群发放信息降到低维表示的方法，以可视化方式分析神经元集群的动态活动。

关键词：神经轨迹；流形学习；降维；神经解码；神经编码

Finding a Low-Dimensional Representation of Neuronal Ensemble Activity

Sung Phil Kim

(Department of Brain and Cognitive Engineering Korea University,
Anam 5ga，SeongbukGu，Seoul，Korea)

Abstract：Large, chronically implanted arrays of microelectrodes are an increasingly common tool for recording from primate cortex and can provide extracellular recordings from many (order of 100) neurons. Neural activity often has dynamics beyond that driven directly by the stimulus. While governed by those dynamics, neural responses may nevertheless unfold differently for nominally identical trials, rendering many traditional analysis methods ineffective. This paper presents a visualization method which reduced the neural cluster dynamic activity down to low-dimensional to analysis dynamic activity of neurons cluster.

Keywords：Nerve orbit，Manifold learning，Dimension reduction，Neural decoding，Neural encoding

1　引言

传统的神经信号分析方法主要是通过分析多次重复实验获得的神经信号，并从中找到一般规律。此方法适用于分析单个神经元在多次实验中的发放特性，如图1左图所示，我们可以用简单的多次实验平均的方法分析单个神经元的活动模式随运动刺激变化的情况。但是当我们关注的是神经元集群在单次实验中随运动刺激变化的发放模式时就无法使用基于多次实验统计的方法，另一方面，直接观察神经元集群的发放又很难看出其变化规律，正如图1右图所示。

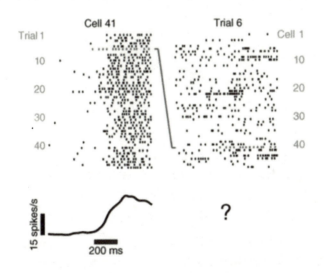

图1　单神经元多次试验分析到多神经元单次试验分析的困难
研究多次重复试验中单个神经元发放模式时（左图），可以用多次试验平均的方法（average across trials）分析其随运动的一般发放模式；而研究单次试验中多个神经元发放模式时（右图），在记录到多神经元发放模式图中，很难观察到与运动相关的一般模式，因此需要新的方法来研究神经元集群的发放模式。

对于多神经元活动的分析需要面对三个挑战：① 如何将神经元集群表示成一个单一实体来分析？ ② 如何统计描述神经元之间的相关性？ ③ 如何利用数据进行建模，并达到数据可视化的目的？

本文的目标是针对运动皮层神经集群活动建立一种低维表征方法，从而能够有效地揭示神经集群活动如何对运动信息进行编码,并可以有效地应用于神经元集群信号的分析。为了实现该目标,以下将对三个关键问题分别进行详细分析:① 如何找到神经元集群的低维表示? ② 神经元集群的低维特征(或者叫做"神经轨迹")能否编码运动参数? ③ 经集群活动的低维特征与运动相关的信息有否损失?

2 神经轨迹

在我们的研究中,我们将神经集群动态活动低维表示为神经轨迹,它代表了神经元集群发放的动态过程。图 2 是 143 个水蛭神经节细胞在水蛭游泳、

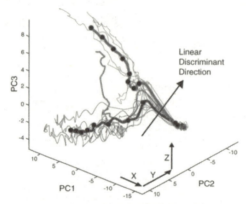

**图 2 143 个水蛭神经节细胞在水蛭游泳、爬行和
徘徊于游泳和爬行之间的运动过程的发放模式的三维可视化**

爬行和徘徊于游泳和爬行之间的运动过程的发放模式的三维可视化[1]。可视化分析所采用的降维算法是主成分分析法(PCA)。图中每条线表示一次实验,靠下面的线表示该过程水蛭处于游泳状态,靠上面的线表示该过程水蛭处于爬行状态,中间一条较粗的线表示一次实验中水蛭初始想游泳而随后改变为爬行的状态。上、下两根粗线代表游泳状态实验或爬行状态实验的平均神经轨迹。从图中可以看出,作为神经元集群的低维表征,神经轨迹方法可以区分出不同的运动模式,并且可以直观地表现出神经集群的动态变化。图 3 是运用神经轨迹应用于神经元集群动态活动分析的另外一个例子,它是 110 个蝗虫触角叶细胞在不同浓度乙醇刺激下的发放模式的三维可视化,使用的降维方法是局部线性嵌入(LLE)[2]。从图中可以看出不同气味强度的刺激可以使神经轨迹呈现出不同的动态过程。图 4 是猴子运动前区的 61 个神经元在 center-out 实验过程中的神经轨迹,该图使用了高斯过程因子分析方法进行

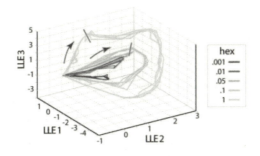

图3　110 个蝗虫触角叶细胞在不同浓度乙醇刺激下的发放模式的三维可视化

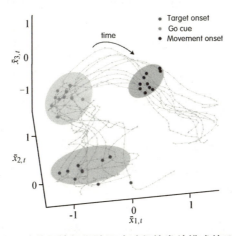

图4　猴子运动前区神经元随运动过程的发放模式的三维可视化

降维[3]。图中下方的点、左上方的点和右上方的点分别代表每次实验中，目标出现、伸抓提示以及运动开始三个运动阶段下神经元集群的发放模式。从图中可以看出，每次实验的神经轨迹都呈现出一定的规律，并且表征三个不同运动阶段的神经轨迹分别聚集在不同区域。以上研究结果表明，神经轨迹可以很好地揭示神经元集群发放活动的内在规律。

以上介绍的例子均采用非监督的降维算法得到神经元集群的低维表示（神经轨迹）。为什么要选用非监督算法呢？主要原因有两个：① 我们要构建的是能够表示神经元集群内在时空结构的神经轨迹，如果使用监督算法就会选择性地保留一些与目标相关的信息而损失掉一些与目标不太相关但可能是有意义的信息；② 最终的解码过程只需要处理神经活动而不处理目标。非监督降维方法有很多种，本文中我们选用的等距特征映射算法（Isomap）是一种非线性方法。基本的 Isomap 通过计算两两样本点之间的测地距离将数据的非线性分布投射到低维空间，如图 5 所示。空间-时间等距特征映射（Spatio-Temporal Isomap）是 Isomap 的一种改进形式[4]，它通过同时计算不同样本间

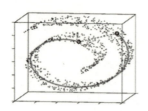

图 5　等距特征映射(Isomap)算法示意图

的空间距离和时间距离来区分不同的样本。对于我们的数据,假设神经活动的信息不仅承载于孤立时间点上的神经发放模式,还与神经发放的时间模式有关系,因此我们选用 Spatio-Temporal Isomap 作为计算神经轨迹的非监督降维算法。图 6 是 ST-Isomap 的一个示意图,可以看到由于它同时计算样本的时空距离,所以它能保持空间变化剧烈的邻近样本的关系。

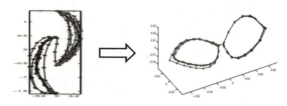

图 6　Isomap 非线性降维算法的一种空间-时间改进算法

3　实验结果

本实验用到的数据是猴子运动前区 61 个神经元的锋电位信号。如图 7 所示,我们使用的实验范式是 Center-Out-Back:猴子手伸向目标然后返回中心,并为下一个目标做准备,共有 8 个方向的目标,每个目标实验 8 次,总共有

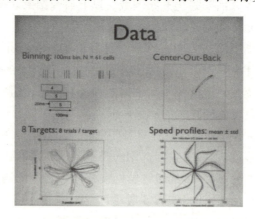

图 7　实验范式和数据示意图

64 个实验。我们采用 100 毫秒时间窗对神经发放加窗,使离散的锋电位发放时间信号变成连续信号。我们同时记录了猴子手的速度信息,如图 7 右下角子图,从图中可以看到手的速度轮廓线分明,同时可以发现速度的变化与目标的位置是独立的。

　　我们将降维后的神经轨迹用于运动解码,并以解码效果作为降维算法模型的评价标准。通过将 ST-Isomap 与未经改进的 Isomap、主成分分析(PCA)、因子分析(factor analysis)等一些常见的降维方法做了比较,实验结果发现 ST-isomap 的解码效果最好(如图 8 所示)。因此,ST-Isomap 为运动的神经解码提供了一个更好的选择。

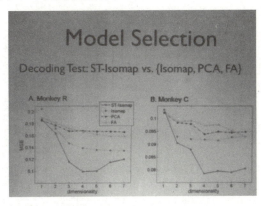

图 8　降维算法比较结果

算法的评价标准是解码预测的运动轨迹与实际运动轨迹的均方差(MSE)

　　找到了神经数据的低维表示(神经轨迹)之后,这些低维的表示反映了运动的什么信息呢? 图 9 以两次相同目标的实验为例,两次实验都采集了 61 个

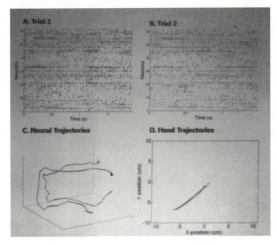

图 9　两个相似运动过程的神经元集群发放模式的神经轨迹依然相似

神经元的活动(如图 9A、B),但是从光栅图中很难看出两次实验中神经发放的共同模式。然而,当它们降维到三维空间时,我们很容易看出在整个运动过程中两次实验中神经活动非常相似。所以,可以说低维表示是神经元集群活动信息的压缩形式的表示。

  图 10 中 A 图显示了所有 64 次实验的神经活动的 3 维轨迹。每一条细线代表一次实验的神经轨迹,红色球代表运动的起始,粗线代表对应每个方向目标的所有实验的平均轨迹。图 B 是将每次实验轨迹按对应目标标记颜色,从图中可以看出相同目标的神经轨迹相对聚集,这种聚集性说明神经轨迹带有目标方位的信息。图 C 中左下方的点代表运动开始,上方的点代表达到最大速度,而右下方的点代表运动终止。

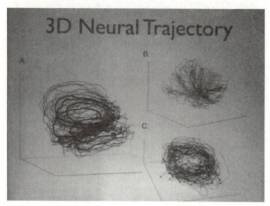

图 10　三维神经轨迹

  神经元集群发放活动的低维表示的另一个好处是:我们可以观察到不同次实验之间的变异性。图 11 显示了若干次方向目标相同的运动过程中神经轨迹之间存在着差异。A 图显示了某一次实验中运动轨迹及其对应的神经轨

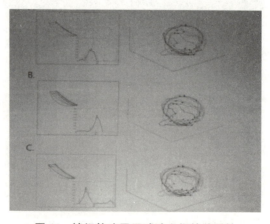

图 11　神经轨迹显示试验之间的差异性

迹;B 图显示了某次实验中猴子的运动开始得较晚;C 图显示了某次实验中猴子在运动早期动错了方向,但随后它又改变到正确的轨迹上来,三次实验相对性的神经轨迹表现出了明显的差异。

神经元集群发放的低维表示(神经轨迹)还能揭示运动编码过程中主要编码的是哪一个运动参数。如图 12 所示,A 图是四次实验猴子手的运动轨迹,B 图是它们对应的神经轨迹。例如,A 图中较下方的两条轨迹具有相同的运动方向而位置不同;在 B 图中它们具有相似的神经轨迹。结果说明了在以上实验中,低维的神经轨迹编码的运动参数是速度而不是位置。

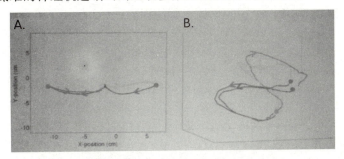

图 12　神经轨迹揭示运动编码过程

以上结果表明,神经轨迹能够反映神经元集群发放的内在规律以及揭示运动编码的信息,那么降维后的低维信息能否用来做运动解码呢? 图 13 给出了一个运用降维信息做运动解码的框架。从图 10 中我们可以看出运动方向相近的数据点彼此在低维空间中位置也相近,因此解码时运动方向可以从训练样本中邻近的样本点得到,而无需建立一个假设的确定性模型(如余弦调制模型)。

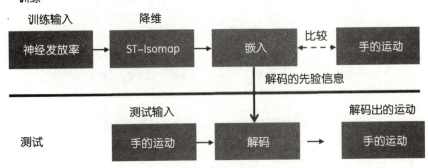

图 13　基于神经轨迹的解码框架

因此,对于方向的解码我们设计了一个概率 K-邻近算法,如式(1)所示。

$$\text{Dir} = \sum_c p_c d_c \tag{1}$$

其中,d_c 为类别 c 的方向,p_c 为 c 方向的概率。

$$\text{Prob}(C = c \mid x) = p_c = (\#NNw/c)/K \tag{2}$$

其中,$(\#NNw/c)$表示 x 的 k 近邻中属于 c 类的根本数。

图 14 显示,在低维流形空间邻近的样本的速度相似,并且速度的大小与低维流形的关系是非线性的。因此,对于速度大小的解码我们采用广义回归神经网络(RBF 神经网络的变种)。整个解码部分主要分为两个过程,首先使用概率模型解码运动方向,然后将使用各个方向的速度模型解码出来的速度用解码出的方向概率加权来解码最终的速度。图 15 展示了我们的解码模型与线性滤波器(LF)、卡尔曼滤波器(KF)以及多层感知机(MLP)神经网络的解码结果比较。我们的模型明显优于其他模型。

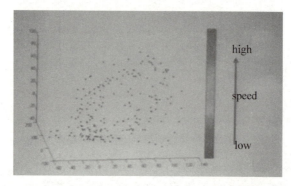

图 14　速度信息在低维空间是局部聚集的并且是非线性的

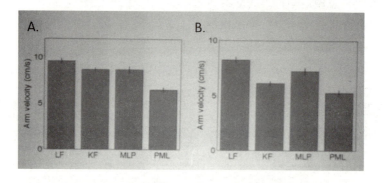

图 15　本文解码算法与一些常用算法解码效果比较

4　总结

ST-Isomap 生成了一个保持神经元集群活动时空发放模式的低维表示。生成的低维运动的神经轨迹代表了伸手的方向和速度。从神经轨迹解码手的速度比从原始神经发放解码效果好。

参考文献

[1] Briggman KL，Abarbanel HDI and Kristan Jr WB. Optical Imaging of Neuronal Populations During Decision-Making. *Science*，2005，307(5711)：896 - 901. DOI：10. 1126/science. 1103736.

[2] Stopfer M，Jayaraman V and Laurent G. Intensity versus Identity Coding in an Olfactory System. *Neuron*，2003，39(6)：991 - 1004

[3] Yu BM，Cunningham JP，Santhanam G. Gaussian-Process Factor Analysis for Low-Dimensional Single-Trial Analysis of Neural Population Activity. *AJP - JN Physiol*，2009，102(1)：614 - 635

[4] Jenkins OC，Matarić CJ. A spatio-temporal extension to Isomap nonlinear dimension reduction. ICML'04 Proceedings of the twenty-first international conference on Machine learning.

讲座人简介

Sung Phil Kim received the B. S. degree in nuclear engineering from Seoul National University，Seoul，Korea，in 1994，and the M. S. and Ph. D. degrees in electrical and computer engineering from Universityof Florida，Gainesville，in 2000 and 2005，respectively. He is an Assistant Professor at the Department of Brain and Cognitive Engineering，Korea University，Seoul，Korea. He was a Postdoctoral Research Associate in Computer Science at Brown University，Providence，RI，until 2009. His current research isfocused on neural decoding for brain-computer interfaces and statistical signalprocessing of neural activity.

图书在版编目(CIP)数据

神经信息工程研究前沿/郑筱祥主编.—杭州:浙江大
学出版社,2012.3
　ISBN 978-7-308-09375-0

　I.①神… Ⅱ.①郑… Ⅲ.①神经科学—信息工程
Ⅳ.①Q189②G202

中国版本图书馆 CIP 数据核字(2011)第 249207 号

神经信息工程研究前沿

郑筱祥　主编

丛书策划	阮海潮(ruanhc@zju.edu.cn)	
责任编辑	阮海潮	
封面设计	姚燕鸣	
出版发行	浙江大学出版社	
	(杭州市天目山路 148 号　邮政编码 310007)	
	(网址:http://www.zjupress.com)	
排　　版	杭州大漠照排印刷有限公司	
印　　刷	临安市曙光印务有限公司	
开　　本	710mm×1000mm　1/16	
印　　张	16	
字　　数	310 千	
版 印 次	2012 年 3 月第 1 版　2012 年 3 月第 1 次印刷	
书　　号	ISBN 978-7-308-09375-0	
定　　价	70.00 元	